Medical Device Cybersecurity for Engineers and Manufacturers

This is a must-read cybersecurity book for anyone learning to engineer, regulate, sell, buy, or maintain medical devices.

—Prof. Kevin Fu, *University of Michigan*
Founder, Chief Scientist, Archimedes Center for Healthcare and Device Security
Credited as the grandfather of medical device security

Medical Device Cybersecurity for Engineers and Manufacturers corrals an expansive problem into a foundational text that will serve generations of engineers very well.

—Seth Carmody, *VP Regulatory Strategy, MedCrypt*

With their great expertise the authors are able to present the cybersecurity of medical devices as a joint task of all involved stakeholders over the entire product lifecycle. This book is a must-read for everyone who is involved in this topic.

—Thorsten Prinz, Ph.D.
Senior Manager, VDE Medical Devices and Software
VDE Association for Electrical,
Electronic & Information Technologies

For a complete listing of titles in the
Artech House Information Security and Privacy Series,
turn to the back of this book.

Medical Device Cybersecurity for Engineers and Manufacturers

Axel Wirth
Christopher Gates
Jason Smith

ARTECH
HOUSE

BOSTON | LONDON
artechhouse.com

Library of Congress Cataloging-in-Publication Data
A catalog record for this book is available from the U.S. Library of Congress.

British Library Cataloguing in Publication Data
A catalogue record for this book is available from the British Library.

Cover design by John Gomes

ISBN 13: 978-1-63081-815-9

Artech House
685 Canton Street
Norwood, MA 02062

Christopher Gates photo © Sabrina Casas Studio, LLC. Jason Smith photo © Lancia E. Smith.

10 9 8 7 6 5 4 3 2 1

Contents

Foreword

Whether we're in the doctor's office for a routine checkup, monitoring our heart rate with a wearable device on a morning jog, or having surgery in the operating room, we rarely think about the devices that we use for medical care being vulnerable to a cybersecurity attack. However, the benefits from rapid advances in the use and capabilities of connected health also come with potential risks.

According to the American Hospital Association (AHA), the United States is home to over 6,000 hospitals, ranging from small community hospitals with less than 50 beds to health systems with over 5,000 beds. In general, hospitals have 3–6 networked medical devices per bed. That means with a total of 931,203 staffed hospital beds across the United States, there are some 3–6 million connected medical devices at hospital bedsides alone, all of which must be protected against cyberattacks. In addition, health delivery organizations, including hospitals, depend on nonmedical devices such as infrastructure systems that keep elevators running, maintain temperature and air quality, and provide lighting and facilitate rapid remote communications for their continued operations. Thus, it is evident that patient safety and quality of care depend on cybersafety, yet we have a large risk exposure of an estimated 10–15 million total connected devices (including infrastructure and supporting devices) that, if compromised, could lead to patient harm or care delivery impact.

This fact is acutely on the minds of both healthcare providers who manage the devices and medical technology and health IT companies who manufacture them. Certain principles must be understood when we think about healthcare cybersecurity:

1. Because the threat landscape is constantly evolving, network and device security have difficulty keeping up;

2. Healthcare institutions do not have the time, money, capabilities, or resources to independently fix cyber vulnerabilities;

3. Patching updates and addressing vulnerabilities in the active medical device ecosystem can be more complicated than your average IT update because there may be a human, not an app or machine, connected to that device and system reboot may not be an option;

4. There are limits to the ability of government regulation to achieve the necessary balance of innovation, effectiveness, security, and privacy.

It is now recognized that these shared challenges are also a shared responsibility. In response, the public and private sectors are working together to address these healthcare cybersecurity challenges in multiple forward-looking ways. It can indeed be argued that working together is an imperative. Healthcare is considered one of the nation's 16 critical infrastructure sectors—like electricity, telecommunications, water, chemicals, transportation, and more. And because private industry owns and operates the vast majority of these critical infrastructures, which constitute essential public resources, it is their responsibility to serve the public good in addition to their commercial interests. That's why this shared responsibility is so important, from the engineers to the C-Suite, to work toward strengthening the security of our healthcare systems and the medical technology that has become indispensable for our nation's health.

The Health Sector Coordinating Council (HSCC)—a public-private partnership of health sector and government stakeholders dedicated to strengthening the nation's critical healthcare infrastructure against all hazards—convenes these interdependent stakeholders to improve the security and resiliency across the healthcare ecosystem. An HSCC task group addressed the medical device security issue head-on, working over 18 months to publish in January 2019 a best practices guide for medical technology companies—the Medical Device and Health I.T. Joint Security Plan (JSP). This is one of the many useful resources referenced in the book you're now reading—*Medical Device Cybersecurity for Engineers and Manufacturers*.

The JSP utilizes security by design principles throughout the product lifecycle of medical devices and health IT solutions. It encourages shared responsibility in the adoption of security-related standards, risk assessment methodologies, and vulnerability reporting requirements to improve information sharing between manufacturers and healthcare organizations. The JSP is a living document and will be updated as appropriate to adapt to the ever-changing threat environment for medical devices and health IT solutions.

More follow-on thought is now being given to how hospitals, device manufacturers, and government can coordinate to better communicate with patients about device vulnerabilities and security, and how we can deal with the challenging issue of aging medical technologies that have reached the end of supported life—whether because of security or operational efficiency—but are not easily replaced because of tremendous expense. Again, the shared challenge meets shared responsibility and collaborative solutions.

These and other critical issues are being addressed in sector-wide work streams, such as

- An HSCC resource for health providers called the Health Industry Cybersecurity Practices (HICP);
- The FDA's September 10, 2019 Patient Engagement Advisory Committee, which sought to understand how best to communicate cybersecurity risks in health risk communications to patients (https://www.fda.gov/advisory-committees/patient-engagement-advisory-committee/september-10-2019-patient-engagement-advisory-committee-meeting-announcement-09102019-09102019);
- Efforts to define and operationalize the imperative for software bills of materials to help health systems understand which software components are in the devices and systems they purchase and hence how to manage associated risk (https://www.ntia.doc.gov/files/ntia/publications/ntia_healthcare_poc_report_final_draft_2019_0904.pdf);
- Expansion of the annual DefCon Biohacking Village Device Hacking Lab, where hackers, healthcare providers, and device manufacturers collaborate to identify vulnerabilities (https://www.villageb.io/device-lab);
- The International Medical Device Regulators Forum (IMDRF) cybersecurity working group, made up of industry and regulators and co-led by FDA and Health Canada, which is seeking to promote a globally harmonized approach to medical device cybersecurity via drafting of a cybersecurity guide (http://www.imdrf.org/consultations/cons-ppmdc.asp).

We see all these collaborations as signs of significant progress. In 2017, a health care cybersecurity task force of industry and government leaders diagnosed that healthcare cybersecurity is in critical condition. While we have had no reports to date that cyber incidents involving medical devices have led to direct patient harm, our collective action, premised on the recognition that patient safety depends on cybersafety, will go a long way toward maintaining public trust in the security, resiliency, and integrity of the life-saving and life-sustaining devices we depend on, and upgrading our healthcare cybersecurity diagnosis to stable.

Good cyber hygiene is found in the products and systems we design, manufacture and maintain, and in the management of those devices and the clinical systems in which they operate. Active and continuous adoption and refinement of resources, like this guide, will help medical technology companies—their engineers, service departments, and executive management, as well as their health provider customers—step up to this shared responsibility.

Greg Garcia,
Executive Director for the Health Sector
Coordinating Council's Cybersecurity Working Group;
Former Assistant Secretary of Cyber Security and Communications
for the U.S. Department of Homeland Security

Acknowledgments

For unflagging support during the composition of this book, the authors would like to thank their spouses, Peggy Wirth, P.J. Gates, and Joyce Smith; their employers, MedCrypt and Velentium; contributing topic authors who lent the industry-leading expertise described in their bios at the back of the book; Artech House editors David Michelson and Rachel Gibson; and Artech's anonymous peer reviewer(s), whose thorough and unwavering eye greatly clarified the final text.

1

Why Secure Medical Devices?

1.1 The Inspiration for This Book

We were inspired to write this book after a long history of working with and for medical device manufacturers (MDMs), as well as with health care delivery organizations (HDOs). The medical device industry was the first industry to be regulated to produce secure embedded devices. This has left MDMs in the odd position of being trailblazers in an industry where cybersecurity is not necessarily their principal skill set. Secure development has nothing in common with the practice of medicine; instead, it has everything to do with subtlety of design, quality of implementation, the threat landscape, and awareness of the attack vectors utilized by attackers. How to balance these concerns with the creation of a new medical device is the overriding topic of this book.

Approaches for incorporating secure development practices into the development lifecycle have not been taught in traditional education programs until very recently, nor are models or best practices available specific to the medical device environment. This lack of training for all levels of engineers, project managers, and senior leadership is a critical shortfall, preventing MDMs from producing proactively secured devices and preventing them from constructively engaging with their customers and regulators.

This book intends to provide guidance to MDMs as to how to implement a secure medical device lifecycle in a manner that is repeatable, trackable, produces artifacts needed for regulatory submission, and actually improves the security standing of the individual medical devices as well as the larger device ecosystem. This book may also serve HDOs as an informative resource for understanding the security activities and support MDMs should perform and how HDOs and MDMs can collaborate to keep devices secure and users safe.

There are many domestic and international standards for securing medical devices (and IoT in general). However, these guidelines are not harmonized and do not provide sufficient details to actually implement such a program. In other words, they provide the *what* but not the *how*. This book intends to close that gap.

We are aware this is a very dynamic topic and that much of the specific technical content included here could be obsolete in as few as 10 years. Notwithstanding, the principles guiding secure development practices advanced within this book will remain viable and valuable even as technology evolves. Moreover, there is a need to identify a foundational position from which incremental improvements can be made and incorporated into maintaining and improving the secure development lifecycle.

1.2   The Evolution of Cybersecurity in Health Care

Over the past decade, our understanding of cybersecurity risks in the health care industry has been evolving. Health care has been moving from a compliance-driven industry to a more cybersecurity aware one. This is largely due to two factors:

1. Driven by government initiatives and the desire to reduce costs and improve efficiency, health care has increasingly digitalized (i.e., implemented more digital systems) and digitized (i.e., amassed more digital data). This increasing connectivity and integration of systems and devices containing more valuable information have not only created a broader attack surface and more attractive opportunity for malicious cyber actors, but it has also increased our dependency on the availability of these systems.

2. Cyber adversaries, be it driven by financial or political motivation, have become increasingly advanced and their attacks have become more sophisticated, targeted, stealthy, and unfortunately, successful. Their impact on businesses around the globe is significant and is now exceeding one trillion USD by some estimates [1].

Consequently, we are no longer thinking about cybersecurity risks merely in the context of our information systems and the data they hold but now have a more comprehensive view, including the potential impact on patient health, care delivery, hospital operations, and even on health systems at the national level.

Unfortunately, this learning experience came from a number of painful events. Ransomware attacks have resulted in the payment [2] of the demanded ransom or have forced hospitals to go through a complex and costly recovery period [3]. For example, the WannaCry attack on the UK National Health Service (NHS) in May 2017 affected more than 1,000 IT systems and medical devices, resulting in the full or partial shutdown of 81 NHS Trust hospitals, the cancelation of 19,000 patient appointments, and a financial impact of an estimated £92 million [4].

This was followed by the NotPetya pseudo-ransomware (wiper) attack in June, which directly affected only a few health care delivery organizations, but impacted hosted documentation services, the availability of pharmaceuticals and vaccines, as well as global shipping and logistics companies [5].

Both attacks resulted in multibillion-dollar losses across the globe and across multiple industries. Yet we need to recognize that both WannaCry and NotPetya were imperfect malware containing design and execution flaws. In other words, in spite of the damage done, it could have been much worse. And, in neither case did we find out who authored the malware nor what their actual intention was (although theories exist).

1.3   The Unique Role of Medical Devices

As medical devices moved from the early days of beneficial uses of electricity [6] and radiation [7] to a tightly integrated system of systems with complex data flows not only between devices, but also between devices and hospital IT systems, we became aware that these devices were not only more vulnerable and difficult to protect, but also that the security compromise of any of these could result in patient harm or impact on care delivery—in addition to the traditional security concerns around data confidentiality, integrity, and availability.

Industry stakeholders have been aware of the security risks of medical devices since security researchers demonstrated their ability to hack into an implantable cardiac defibrillator (ICD) in 2008 [8]. Since then we have seen individual researchers [9], hospitals [10], and government agencies [11] conduct their own additional research, all with the same results—our medical device ecosystem is highly vulnerable and we have historically failed to establish cybersecurity as a buying criterion and as a design objective. Regulators and leading medical device manufacturers have recognized the problem and are establishing a path forward.

It should also be understood that even though this book is generically using the term medical devices, the lifecycle management and security best practices described can equally be applied to a wide range of devices that are used on hospital networks, not just the ones classified as medical devices by the local regulators. A cyber incident involving a hospital elevator impacts patient transportation, a change in temperature or humidity in the operating theater may force procedures to be delayed, or shutting down a blood or organ refrigerator can have a serious impact on patients waiting for the transplant or infusion. Therefore, even though international regulators' guidance is focused on medical devices and their potential security risks to patient harm or care delivery, the practices outlined herein are equally beneficial and can be applied to other care-critical, albeit not regulated, device types.

Clearly, cybersecurity for medical devices needs to be proactively addressed on the level of the individual device. But since no device will ever be perfectly secure (and certainly not devices already in use today), additional security measures should be applied on the level of the integrated system of devices, be it in the traditional, hospital-based care environment or in the evolving telehealth and home care settings. A system is only as secure as the sum of its parts and can only achieve a sufficient level of security if all stakeholders accept and live up to their responsibilities. However, it has generally been accepted that the security posture of the device itself is the most critical component and that device manufacturers need to live up to their responsibility of providing more secure designs and devices with an easier-to-maintain security posture [12].

Today's medical device ecosystem goes far beyond the device or local device network itself and includes electronic health record (EHR) and image management (PACS) systems, inventory and business automation systems, cloud based value-added services, access via mobile devices, home based routers, and public networks. In order to provide sufficient security, each party needs to understand and accept their responsibility as well as provide their contribution.

In the most general sense, we need to fulfill the four main security tenets: secure the device, manage the device, manage the ecosystem, and respond to incidents. This is outlined in more detail in Table 1.1.

1.4 Regulatory Environment

Over the past years, many international regulators have taken steps to provide a guidance framework and to establish their expectations on cybersecurity for medical devices. The US Food and Drug Administration (FDA) has been taking the lead with its Premarket Guidance (2014) [13], Postmarket Guidance (2016) [14], and updated draft Premarket Guidance (2018) [15]. Other notable contributions have been and are being provided by the International Medical Device Regulators Forum (IMDRF) [16] and the European Union [17], in addition to individual countries like Canada, Japan, China, Australia, and France [18].

Table 1.1
The Four Security Tenets

Secure the Device	
Manufacturer	**Operator**
Software design best practices	Secure handling
Hardened design	Secure media use
Complementary endpoint security	Integration best practices
Cryptographic protection	Secure networking
Manage the Device	
Manufacturer	**Operator**
Lifecycle management	Procurement and contracting
Security V&V	Asset management
Vulnerability disclosure	Dependency management
Software Bill of Materials	Risk assessment and management
Protect the Ecosystem	
Manufacturer	**Operator**
Secure remote access	Network architecture
Strong authentication	Event monitoring
Security documentation	Firewalls/gateways
Enablement and training	Enablement and training
Respond to Incidents	
Manufacturer	**Operator**
Threat monitoring	Detect, respond, recover
Customer support	Forensics
ISAO participation	Communication and decision-making
Regulatory reporting	Reporting as needed

The regulatory background will be further discussed in subsequent chapters, however, a commonly voiced objection has been that this is guidance only and therefore not a legally binding requirement. This is a common misunderstanding in the sense that many regulators see guidance as clarification on the agency's current thinking based on already-existing regulations. In other words, regulators have stated that cybersecurity is no longer optional. The exact details on how to set up a secure lifecycle management process that results in sufficiently secure devices may be the manufacturer's choice, but that fact that such a program is needed is not considered optional.

The health care and specifically the medical device industries are highly regulated to assure patient safety and reliable quality care delivery. Unfortunately, these regulations, combined with a formal design process and conservative decision making, are just the opposite of what we are used to in information technology where we want to be flexible and nimble so as to quickly respond to new cyber threats and risks.

In short, we need to improve our processes to be able to move faster in response to security challenges without compromising device safety. But we also need to improve device design to provide a more secure state out of the factory so that fewer downstream field changes are needed, while also improving device design so that these fewer changes are easier and faster to deploy. This in itself is already a significant challenge for manufacturers and device operators.

The challenge facing the medical device industry is to move to a more secure state without leaving the benefits of controlled processes and deliberate decision-making behind. We hope that this book will contribute to this needed change by providing the how-to of medical device security for device manufacturers and their engineers. (Guidance for health care providers will be provided only on a high level, as detailed content for them has been provided elsewhere [19].) We will touch on security technologies and best practices to a certain extent but will not offer a deep dive. Medical devices are too complex and vary too widely in the use case, risks, and technological capability to allow for a meaningful selection of security tools and technologies in this book. Instead, we will provide a detailed discussion of how to enable a security-capable organization and how to build a process framework to assure security-inclusive lifecycle management for the medical device manufacturing process.

1.5 Looking Ahead

Over the past decade, much research has been published about the potential risks of failing to secure medical devices. We do not have the luxury of not taking this problem seriously. Security researchers have demonstrated over and over again the often-appalling security weaknesses of medical devices and how they could be exploited by an attacker, including examples given in this chapter. But we should not think of this problem only in the context of a targeted attack with malicious intent:

- Since 2015, security researchers have repeatedly gathered evidence that medical devices are used as an entry point and beachhead for an attack on other targets [20].

- Even if the device is not the target at all, it may get caught up in an attack if the device happens to fit the threat profile, as happened during the WannaCry malware outbreak [21].

- In another nontargeted event, medical equipment in a Siberian hospital was shut down by ransomware in the midst of brain surgery. Even though the patient was reported unharmed, this was a close call [22].

- We know that even minutes of delay in care can have impact on emergency patient mortality rates [23] and that in the aftermath of a cyberattack hospitals saw additional patient deaths due to heart attacks [24]. Due to the care-critical nature of medical devices, any security incident affecting device and service availability will indirectly impact patient safety.

We also need to ask ourselves that even though we have no reported case of direct patient harm, have we really been able to correctly attribute impact on patient outcomes to cybersecurity events? Did we look? Did we know where to look? Have we even had the right data and capabilities to detect such an incident?

Even though we need to resist the urge to have this discussion driven by sensationalism and headlines, we also need to proceed with a sense of urgency. We hope that this book will contribute to a measured and reasonable approach that ultimately leads to a more secure and therefore safer health care ecosystem. In the end, it is about preventing patient harm and preserving patient trust.

References

[1]   "Cybercrime Could Cost Companies Trillions over the Next Five Years," HelpNet Security, Jan. 21, 2019, https://www.helpnetsecurity.com/2019/01/21/cybercrime-could-cost-companies-trillions/.

[2]   Winton, R., "Hollywood Hospital Pays $17,000 in Bitcoin to Hackers; FBI Investigating," Los Angeles Times, Feb. 18, 2016, https://www.latimes.com/business/technology/la-me-ln-hollywood-hospital-bitcoin-20160217-story.html.

[3]   Davis, H. L., "ECMC Spent Nearly $10 Million Recovering from Massive Cyberattack," *The Buffalo News*, July 26, 2017, https://buffalonews.com/2017/07/26/cost-ecmc-ransomware-incident-near-10-million/.

[4]   Smart, W., "Lessons Learned Review of the WannaCry Ransomware Cyber Attack," Department of Health & Social Care, NHS England, Feb. 1, 2018, https://www.england.nhs.uk/wp-content/uploads/2018/02/lessons-learned-review-wannacry-ransomware-cyber-attack-cio-review.pdf.

[5]   Greenberg, A., "The Untold Story of NotPetya, the Most Devastating Cyberattack in History," *Wired Magazine*, Aug. 22, 2018, https://www.wired.com/story/notpetya-cyberattack-ukraine-russia-code-crashed-the-world/.

[6]   Waller, A. D., "A Demonstration on Man of Electromotive Changes Accompanying the Heart's Beat," *The Journal of Physiology*, Oct. 1, 1887, https://www.ncbi.nlm.nih.gov/pubmed/16991463/.

[7]   Röntgen, Wi. C., *Über eine neue Art von Strahlen [On a New Kind of Rays]*, Würzburg, DE: Stahel, 1896.

[8]   Halperin, D., T. Heydt-Benjamin, B. Ransford, et al., "Pacemakers and Implantable Cardiac Defibrillators: Software Radio Attacks and Zero-Power Defenses," *2008 IEEE Symposium on Security and Privacy*, Oakland, CA, May 18–24, 2008, pp. 129–142, https://ieeexplore.ieee.org/document/4531149.

[9]   "ICS Alert: Medical Devices Hard-Coded Passwords," DHS CISA, Oct. 29, 2013, https://www.us-cert.gov/ics/alerts/ICS-ALERT-13-164-01.

[10]   Bruemmer, D., and F. Hudson, "Medical Device Security in a Connected World," *11th AMC Conference on Securely Connecting Communities for Improved Health*, Chapel Hill, NC, June 22, 2015, https://nchica.org/wp-content/uploads/2015/06/Bruemmer-Hudson.pdf.

[11]   "FDA and DHS Increase Coordination of Responses to Medical Device Cybersecurity Threats under New Partnership; a Part of the Two Agencies' Broader Effort to Protect Patient Safety," FDA News Releases, Oct. 16, 2018. https://www.fda.gov/news-events/press-announcements/fda-and-dhs-increase-coordination-responses-medical-device-cybersecurity-threats-under-new.

[12]   Van Wagenen, J., "Medical Device Vulnerabilities Continue to Plague the Industry," HealthTech Magazine, Dec. 11, 2018, https://healthtechmagazine.net/article/2018/12/medical-device-vulnerabilities-continue-plague-industry.

[13]   "Content of Premarket Submissions for Management of Cybersecurity in Medical Devices; Guidance for Industry and Food and Drug Administration Staff," FDA Guidance Documents, Oct. 2, 2014, https://www.fda.gov/regulatory-information/search-fda-guidance-documents/content-premarket-submissions-management-cybersecurity-medical-devices-0.

[14]   "Postmarket Management of Cybersecurity in Medical Devices: Guidance for Industry and Food and Drug Administration Staff," FDA Guidance Documents, Dec. 28, 2016, https://www.fda.gov/regulatory-information/search-fda-guidance-documents/postmarket-management-cybersecurity-medical-devices.

[15]   "Content of Premarket Submissions for Management of Cybersecurity in Medical Devices: Draft Guidance for Industry and Food and Drug Administration Staff," FDA Guidance Documents, Oct. 18, 2018, https://www.fda.gov/regulatory-information/search-fda-guidance-documents/content-premarket-submissions-management-cybersecurity-medical-devices.

[16]   "Principles and Practices for Medical Device Cybersecurity," IMDRF, Dec. 2, 2019, http://www.imdrf.org/consultations/cons-ppmdc.asp.

[17]   "MDCG 2019-16–Guidance on Cybersecurity for Medical Devices," MDCG European Commission, Dec. 2019, https://ec.europa.eu/docsroom/documents/38941.

[18]   Morgan, C., "Medical Device Cybersecurity Regulatory Publications," Apraciti, Nov. 26, 2019, https://www.apraciti.com/blog/2019/11/25/global-regulatory-authority-publications-on-medical-device-cybersecurity.

[19]   Wirth, A., and G. Stephen (eds.), Medical Device Cybersecurity: A Guide for HTM Professionals, Association for the Advancement of Medical Instrumentation (AAMI), May 2018, https://my.aami.org/store/SearchResults.aspx?searchterm=mdc&searchoption=ALL

[20]   TrapX Security, "Anatomy of an Attack – Medical Device Hijack (MEDJACK)," May 7, 2015, https://trapx.com/trapx-labs-report-anatomy-of-attack-medical-device-hijack-medjack/.

[21]   Brewster, T., "Medical Devices Hit by Ransomware for the First Time in US Hospitals," Forbes, May 17, 2017, https://www.forbes.com/sites/thomasbrewster/2017/05/17/wannacry-ransomware-hit-real-medical-devices/#741c2b03425c.

[22]   Pitalev, I., "Russian Hackers Switched Off Medical Equipment During Kid's Brain Surgery," Russia Beyond, Jul. 6, 2018, https://www.rbth.com/lifestyle/328711-russian-hackers-switched-off.

[23] Jena, A., N. Mann, L. Wedlund, and A. Olenski, "Delays in Emergency Care and Mortality during Major U.S. Marathons," *New England Journal of Medicine*, Apr. 13, 2017, https://www.nejm.org/doi/full/10.1056/NEJMsa1614073.

[24] Holmes, A., "The Rise of Cyberattacks and Data Breaches Against US Hospitals has been Linked to an Uptick in Heart Attack Deaths," *Business Insider*, Nov. 13, 2019, https://www.businessinsider.com/cyber-attacks-hospitals-rise-in-heart-attack-deaths-study-2019-11.

2

Establishing a Cybersecurity Focus

This book will present the key elements of a lifecycle-based approach for an MDM's device cybersecurity program. Good cybersecurity practices, especially in a regulated, quality-focused, and tightly controlled engineering and manufacturing environment as seen in the healthcare industry, always require an approach that combines technology, process, organization, communication, and culture.

We hope that this book will be useful to the individual reader across a spectrum of roles and will help manufacturers to build or improve their cybersecurity programs. In this chapter, we will lay the groundwork to help companies, whether at the beginning of their journey or already on their way, to develop and refine their approach and organizational capability. We will provide an overview of the foundational steps: governance and organization, regulatory background, supply chain, and general practices required to be able to deliver secure, and therefore safe, medical devices to the market. These individual elements will then be analyzed in greater detail in the remainder of this book.

Although understanding the international regulatory environment is required for any MDM [1], we will define a lifecycle-based approach to cybersecurity that is largely independent of local regulations and can be applied no matter where your devices are developed, manufactured, and marketed. We will, however, use the FDA's cybersecurity guidances as the general reference point due to the FDA's leadership in medical device cybersecurity [2] as well as the fact that the United States is the single largest medical device market in the world and will therefore likely affect the vast majority of manufacturers.

2.1   Security Governance

One of the more critical challenges facing medical device manufacturers and their customers today is to ensure that all products and solutions can withstand cyberattacks. With a flourishing digital revolution and connected healthcare ecosystem, it is clear that to meet these challenges manufacturers must take a strategic and integrated view of product security and establish a comprehensive risk-based medical device cybersecurity program with strong leadership and governance.

The proliferation of millions of connected digital devices, in the traditional hospital environment and the evolving care delivery space like patients' homes, allows users and healthcare providers to share, search, navigate, manage, compare, and analyze a virtually limitless flow of data that can be used to enhance care outcomes. This digital ecosystem has already helped the industry expand the portfolio of personal and health care oriented smart devices, sparked innovation, and increased service efficiency.

However, the exponential increase in the volume and types of data available also leads to increased vulnerability to cybercrime—healthcare data is the #1 target for cybercriminals due to its content, opportunities to monetize in many different ways, and perceived low defenses of health care organizations. These threats include malicious security attacks via viruses, worms, and hacker intrusions. Governments around the world have enacted legislation to criminalize many of these cyberattacks and to assure the confidentiality, integrity, and availability of protected health information (PHI) and personally identifiable information (PII) (e.g., US-HIPAA, Canada-PIPEDA, general privacy legislation under the European Directive 95/46/EC, Japan-PIPA, and others).

Some sensitive data areas that present as particularly vulnerable are:

- Provider networks;
- Personal health devices;
- Remote services;
- Sensitive data storage;
- Sensitive data on the move.

Tracking and addressing security concerns requires constant vigilance. Recently, security researchers disclosed 11 different zero-day vulnerabilities within Wind River's VxWorks [3], a real-time operating system used in over two billion embedded systems that include medical devices, routers, VOIP phones, and mission-critical infrastructure equipment. Successful exploitation of this vulnerability could allow an unauthorized user to execute arbitrary code on the target system. The unauthorized user could then install programs; view, change, or delete data; or create new accounts with full user rights.

January 2020 saw health care organization prepare for the expiration of Microsoft's Extended Support period for Windows 7 and Windows Server 2008 R2 (Extended Security Updates (ESU) for critical and important issues can be purchased for a maximum of 3 years after that). Only a broadly coordinated review-and-replace program for legacy systems, including product-specific verification, validation, authorization, and communication, would mitigate any serious continuity issues and security concerns.

These are just two examples of the constant barrage of challenges facing medical device manufacturers today.

2.1.1 Effective Oversight

To address complex and growing best-practice security needs, as well as regulatory and legal compliance requirements, internal entities charged with managing product security must also be charged with designing and monitoring mitigation structures and strategies. This entails creating policies, procedures, and processes for safe and effective deployment of technology solutions. Additionally, this requirement calls for notification and management of incident response through monitoring of the deployed technology solutions. In short, it requires effective governance.

The goal of effective governance is to enable the organization to manage product security risk end-to-end across all stages of the product development and maintenance lifecycle, to proactively design and implement capabilities into product solutions and sustaining operations that minimize security risk exposure, and to mitigate and/or remediate premarket and postmarket product security risk exposures that may occur. A connected, interoperable health care ecosystem demands an integrated view of product security and an unwavering adherence to strategic security initiatives.

An in-house security governance team acts in a continuous leadership and oversight capacity to develop and execute strategies and structures that successfully implement the critical attributes of a product security program including policies, risk assessments, security testing, communications, stakeholder requirements, incident management, metrics, and a maturity road map for continuous improvement. The team also coordinates the efforts of external players across the cybersecurity ecosystem (customers, vendors, regulators, standards organizations, industry groups, and researchers, among others) through ongoing dialogue. This is extremely productive in building key relationships and promoting industry best practices toward the safety and security of personal and medical devices. Critically, alignment of executive leadership within the organization secures the necessary buy-in for the team to move forward successfully.

The governance team's goals and motivation should be to enable a manufacturer's organization to deliver on key cybersecurity capabilities:

- Prevent unauthorized access of medical devices and patient information;
- Prevent compromise or loss of patient data;
- Ensure medical device functional integrity and services availability to enable safe and reliable patient care;
- Enable up-to-date security patching to remediate unsecure systems and vulnerabilities;
- Integrate security compliance controls into product software development processes.

Within the team it is important to establish roles and responsibilities across the risk management process to deal with regulatory entities and assure that security standards meet or exceed current regulatory requirements, industry best practices, and customer needs.

The team must promote the consistent adoption of strategies to proactively address risks and threats, including what are often referred to in the area of cybersecurity as The Three Deadly Sins.

- *Password risk:* the risk from a lack of strong identity and permission management (e.g., mitigate via multifactor authentication);
- *Encryption risk:* the risk from a lack of strong end-to-end data encryption—from the source where data is generated, over the network, and when resting in a data center—and/or effective data-loss prevention solutions;
- *Patch management risk:* the risk from a lack of effective patch management, failing to timely address vulnerabilities in, for example, legacy operating systems.

A unified effort to deliver on key security tenets is essential to enable safe and secure products and services as well as to reduce potential exposures to data breaches, third party vulnerabilities, and sanctions from regulatory institutions and customers. By building capabilities from within the organization, a security governance team develops and maintains the skills necessary to successfully manage this effort—skills that include organizational enablement, global monitoring, case escalation, rapid response, and full management visibility to security issues.

2.2 Building a Security-Capable Organization

Health care providers and patients need assurance that the technology they interact with on a daily basis is as secure as possible. A security-capable medical device manufacturer focuses its energies on advancing product security across its portfolio. Key actions and initiatives are defined, mapped against a timeline, and executed with conviction.

A security-capable medical device manufacturer addresses the evolving nature of security, including consistent review of product feature requirements, security threat assessments, tracking and monitoring, and compliance with federal/local government regulatory standards. By creating an internal cybersecurity program, the manufacturer becomes proactive rather than reactive.

As shown in Figure 2.1, the five basic tenets of a robust medical device cybersecurity program are:

- Strong governance;
- Ongoing testing;
- Coordinated vulnerability disclosure;
- Software bill of materials (SBOM);
- A road map forward.

2.2.1 Strong Governance

It is important to convert an area of potential concern into a knowledge-sharing engagement opportunity that will help refine critical thinking and lead to development of solutions that enable regulatory compliance, as well as result in more secure and therefore safer products. As described previously (Figure 2.1), qualified governance of a comprehensive risk-based security program can guide a capable organization through the challenges of today's cybersecurity environ-

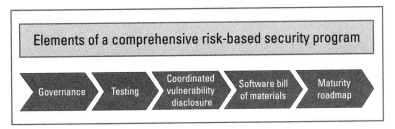

Figure 2.1 Elements of a comprehensive risk-based security program.

ment. The team must develop and deploy strict standards ubiquitously across all systems.

2.2.2 Ongoing Testing

The objective of ongoing penetration testing and static code analysis is to achieve early detection and monitoring of vulnerabilities and to develop a response plan as necessary. The priority is always to mitigate any situation and develop a solution or workaround. If this process is conducted in-house by a dedicated team of ethical hackers, it can be a fast and cost-effective opportunity to secure products, when compared to employing more expensive external resources.

Product and services security testing should cover a wide variety of cybersecurity tasks, including:

- Security vulnerability and penetration testing;
- Security risk assessments;
- Security source code analysis;
- Third party vendor engagements;
- US Department of Defense (DoD) technical product security testing;
- Security training tailored to unique roles including product architecture, development, and testing;
- Tool validation;
- Tool evaluation;
- Threat monitoring;
- Metrics for product development.

Internal product security awareness across an organization can also be heightened with initiatives such as hacking events (such as those hosted during the RSA and DEFCON conferences), security workshops, and social media announcements that cover the latest security updates. As an example, Philips established a Security Center of Excellence (SCoE), recently certified under UL IEC 62304 in 2015 to develop products that are cyber resilient. At the SCoE, a dedicated team of ethical hackers, or security ninjas, engages in continuous vulnerability and penetration testing to proactively identify product weaknesses. Complementing and strengthening the product security testing of Philips product engineering and development teams, the SCoE testing processes and results are defined in standardized use-case scenarios for a common-response approach, which are then leveraged across the entire Philips global enterprise and integrated into risk assessment, secure development lifecycle (SDLC), and maintenance procedures.

2.2.3 Coordinated Vulnerability Disclosure

The development of a coordinated vulnerability disclosure program is required to reassure customers that proper effort will be made to quickly mitigate any vulnerabilities to minimize damage potential. It is important to handle all vulnerability disclosures, whether reported through security researchers or based on a vulnerability discovered as the result of an actual security incident, with a sense of urgency and sensitivity. A formal incident response management process must be put into place, which includes documenting all communication, opening a corrective action program, developing a solution, and authoring an incident report.

Confirmed vulnerabilities must result in a direct report into government agencies such as the U.S. DHS (ICS-CERT program) and then communication through the press to the public. The FDA's Pre and Postmarket Management of Cybersecurity in Medical Devices guidance documents, published in October 2014 and December 2016, respectively, provide direction on key principals that are globally applicable in practice and in cooperation with other governmental entities and processes. Transparency is key.

Over the past few years, several leading medical device manufacturers have designed and implemented a Coordinated Vulnerability Disclosure Policy and as a result have been globally recognized as industry leaders with fully developed and operationally mature processes behind their policy. When public media attention is drawn to security incidents, it is these manufacturers who are often singled out as the ones prepared and capable to address difficult security issues.

2.2.4 SBOM: Commercial and Open-Source Software Governance

Many organizations do not have an accurate listing of the SBOM of their products. Since there is no accurate listing, they have no idea of the vulnerabilities associated with the various components. When faced with a security weakness, they cannot identify the affected code and introduce a solution. Hence, an agile response is impossible.

Organizations reliant on integration of third-party software leave themselves open to hidden risks posed by programming code that is not their own. To prepare for pending legislation on this topic globally, creation of a SBOM for every product is essential. This identifies and describes open source or commercial third-party software components and allows organizations to quickly identify and respond to possible security vulnerabilities. The Cyber Supply Chain Management and Transparency Act will require U.S. government agencies to obtain SBOMs for any new products they purchase, including medical devices. It will require obtaining SBOMs for any software, firmware, or products containing a third party or open source binary component [4].

A carefully defined 3-step approach to commercial and open source software governance and compliance is essential to address this challenge and includes:

- *Deploy:* Generate SBOM on all software products developed. This is accomplished by deploying survey tools across all business groups involved.

- *Integrate:* Inspect the source code and/or binaries of the product, then integrate with the software build server to generate the bill of materials.

- *Report:* Create a security risk summary of each product, correlated with the known security vulnerabilities associated with the components identified.

As medical device manufacturers build their systems, they must look at checkpoints along the way, testing and harmonizing security aspects as they go. The SBOM should be integrated into the software development lifecycle for every product, including updating and maintenance procedures. SBOM vulnerabilities and license issues must be identified and the findings incorporated into a security risk assessment with risk remediation. SBOMs must be monitored continuously for new vulnerabilities and security software updates. Relevant product documentation must then be revised to reflect the very latest SBOM.

2.2.5   Maturity Road Map

Integrating product security into new product development and consistently deploying product security processes across a portfolio sets the stage for a manageable future. The purpose and intent of a maturity road map is to measure and improve processes and organizational capabilities. The ultimate goal is to attain improved levels of product security maturity with new product introductions, ongoing service operations, and postmarket lifecycle management. It begins with an assessment and monitoring of installed base/legacy products to detect OS obsolescence, incompatibilities, and hardware/firmware vulnerabilities, then allows for ongoing, timely maintenance/updating and lifecycle scheduling.

2.2.6   Security Designed In

The concept of security designed in (that is end-to-end, from design to production to support) is key to the long-term success of an organization's products, services and solutions. Infusing security principles begins with product design and development, through testing and deployment—and is followed up with robust policies and procedures for monitoring, effective updates, and when necessary, incident response management.

Engineers armed with data on potential security weaknesses can define configuration changes and re-engineering efforts that will harden a system against outside threats. The same information will also drive security design requirements for new products.

2.2.7   Section Summary

The successful deployment of secure products and services can be achieved by being diligent in the application of comprehensive policies and procedures, including the establishment of a long-term strategy for lifecycle management of obsolete products. Transparency, accountability, and responsiveness must be ongoing commitments as organizations maintain and evolve their cybersecurity programs—giving customers, businesses, and employees the protection they are entitled to amid growing global cybersecurity threats.

Frequent dialogue between medical device manufacturers, hospitals, regulators, and security professionals—particularly around interoperability—will advance innovations in security and the healthcare industry.

2.3   Regulations and Standards

The medical device industry is highly regulated. This is necessary to ensure that these products are safe and effective for patients and users. Regulators put a strong focus on a comprehensive quality system, including robust risk management and design controls. In any highly regulated industry, standards play an important role to help guide the necessary processes and controls needed for safe and effective products. Standards help the industry establish a basis for mutual understanding and help to serve as tools to facilitate manufacturing, design, risk management, and other essential elements of quality. Regulators, in turn, depend on these standards to set the bar on process and organizational activities when assessing and guiding manufacturers. As medical device manufacturers consider their approach for addressing cybersecurity, it is therefore important that they include regulatory expectations and appropriate standards to inform their process.

Health care is considered a critical infrastructure in many countries, including the United States, and cybersecurity has emerged as a real risk to this and other infrastructures in the past decade. Therefore, the pressure has increased to address cybersecurity as a significant risk not just to privacy and confidentiality but also to patient safety and access to the overall healthcare infrastructure. As a result, regulators have released specific guidance documents in various countries across the globe to better describe their expectations. The industry has also responded with new standards to help guide manufacturers and healthcare providers. Many manufacturers have also found value in more general security standards and best practices, so this chapter will also review a selection of these.

2.3.1   Regulatory Considerations

Navigating cybersecurity expectations as presented by regulators can be a bit tricky. Medical device software is something that the US FDA has provided guidance on since 1997, with such guidance documents as General Principles of Software Validation (version 1.1 issued June 9, 1997) and Guidance for the Content of Premarket Submissions for Software Contained in Medical Devices, issued May 29, 1998. These guidance documents were rereleased in 2002 and 2005, respectively, which is still over a decade ago. The fact that the FDA has provided guidance on how to properly address software in medical devices means that cybersecurity of that software was recognized as a relevant issue. But there are valid reasons that the first FDA draft cybersecurity guidance was not released until 2013. First, cybersecurity risk in general has been building for years and major events have made news in various sectors. Starting in 2006, security researchers in the United States began to focus on certain embedded software updates [5] and in 2008, researchers exposed vulnerabilities in Implantable Cardiac Defibrillators [6]. Then, 2011 became the year of the insulin pump where researchers began to focus on this product line and several peer-reviewed vulnerabilities were published [7]. As a result of these highly publicized vulnerabilities, the Government Accountability Office published a report stating that the "FDA Should Expand Its Consideration of Information Security for Certain Types of Devices" [8].

During this time, the FDA was engaging in discussions with researchers, medical device manufacturers, and hospitals, trying to bring these groups together and provide a shared-responsibility mindset promoting cooperation and communication. FDA released its first cybersecurity-specific guidance in 2014, "Guidance for the Content of Premarket Submissions for Software Contained in Medical Devices." Over the next 5 years, the United States continued to provide guidance in this space and was joined by other countries to provide cybersecurity-specific medical device guidance, including Japan, China, Australia, and Canada (see Table 2.1). Most of these guidance documents stress the viewpoint that cybersecurity is considered part of existing safety and quality expectations already defined in existing regulations, and the guidance document serves as an interpretation of these regulations specifically for cybersecurity. The Medical Device Coordination Group (MDCG) in the EU provided a focused interpretive guidance to assist member states and manufacturers on the security-specific interpretation of clauses in the Medical Device Regulation (MDR) in the EU, outlining what type of cybersecurity documentation

Table 2.1

List of Guidance Documents 2005–2019

Country	Document Title	Link
2005		
United States	Cybersecurity for Networked Medical Devices Containing Off-the-Shelf (OTS) Software	https://www.fda.gov/ regulatory-information/search-fda-guidance-documents/ cybersecurity-networked-medical-devices-containing-shelf-ots-software
2014		
United States	Final Guidance: Content of Premarket Submissions for Management of Cybersecurity in Medical Devices	https://www.fda.gov/ regulatory-information/search-fda-guidance-documents/ content-premarket-submissions-management-cybersecurity-medical-devices-0
2015		
Japan	Ensuring Cybersecurity of Medical Device: PFSB/ELD/OMDE Notification No. 0428-1	Japanese Version - https://www.pmda. go.jp/files/000204891.pdf
2016		
United States	Final Guidance: Postmarket Management of Cybersecurity in Medical Devices	https://www.fda.gov/regulatory-information/search-fda-guidance-documents/postmarket-management-cybersecurity-medical-devices
2017		
China	Medical Device Network Security Registration on Technical Review Guidance Principle	Not available
2018		
Germany	Cyber Security Requirements for Network-Connected Medical Devices	https://www.allianz-fuer-cybersicherheit. de/ACS/DE/_/downloads/BSI-CS/BSI-CS_132E.html?nn=6656412
Japan	Guidance on Ensuring Cybersecurity of Medical Devices: PSEHB/MDED-PSD Notification No. 0724-1	Japanese Version https://www.pmda. go.jp/files/000225277.pdf
Singapore	Information Technology Standards Council Technical Reference 67: Connected/Security Medical device	https://itsc.imda.gov.sg/standards/ singapore-it-standards/
South Korea	Cyber Security Guide for Smart Medical Service	Not available

and process should be considered appropriate for the more general expectations outlined in the MDR.

For the most part, the various regional guidance documents stress the importance of appropriate security controls, risk management, vulnerability monitoring and communication, update and patch management, incident

Table 2.1 (continued)

United States	Draft Guidance: Content of Premarket Submissions for Management of Cybersecurity in Medical Devices	https://www.fda.gov/regulatory-information/search-fda-guidance-documents/content-premarket-submissions-management-cybersecurity-medical-devices
2019		
Australia	Medical device cyber security guidance for industry	https://www.tga.gov.au/publication/medical-device-cyber-security-guidance-industry
Canada	Premarket Requirements for Medical Device Cybersecurity	https://www.canada.ca/en/health-canada/services/drugs-health-products/medical-devices/application-information/guidance-documents/cybersecurity.html
France	Cybersecurity of Medical Devices integrating software during their lifecycle	https://www.ansm.sante.fr/S-informer/Points-d-information-Points-d-information/L-ANSM-lance-une-consultation-publique-sur-un-projet-de-recommandations-pour-la-cybersecurite-des-dispositifs-medicaux-Point-d-information
2020		
Saudi Arabia	Guidance to Pre-Market Cybersecurity of Medical Devices	https://www.sfda.gov.sa/ar/medicaldevices/regulations/DocLib/MDS-G38.pdf
IMDRF	IMDRF Principles and Practices for Medical Device Cybersecurity	http://www.imdrf.org/consultations/cons-ppmdc.asp
European Union	MDCG 2019-16 Guidance on Cybersecurity for medical devices	https://ec.europa.eu/docsroom/documents/38924

response, labeling, testing, and supply chain management. They have also been driving the expectation for specific artifacts such as the SBOM, which is a list of software components within the product. This information allows for the monitoring of new vulnerabilities from the supply chain and helps both vendors and users better understand the risks within their components. There has also been a push from regulators to use the Manufacturers Disclosure Statement for Medical Device Security (MDS²) form, which helps to communicate the security capabilities supported by the medical device.

Threat modeling and security risk management have been two additional topics that have received the regulators' interest, in part because these processes help manufacturers identify their own risks and in part because these documents are particularly essential for the regulators' ability to understand a medical device's risk profile. Good risk management activities are so central to a good product security management process that this is highlighted in every guidance document to date, often referring to AAMI TIR57 Principles for medical device security—risk management as a recommended best practice.

2.3.2 Standards

For many aspects of medical device design and development process, there is a central, foundational standard that outlines the essential elements of that process. For example, for software development in medical devices, manufacturers use IEC 62304 Medical Device Software—Software Life Cycle Processes. Safety risk management is guided by ISO 14971 Medical Device —Application of Risk Management to Medical Devices and the quality management system is constructed by following ISO 13485 Medical Devices—Quality Management Systems—Requirements for Regulatory Purposes. When a manufacturer conforms to one of these international standards, it gives them the confidence that they have met the basic requirements for that aspect of the process. When seeking to meet the necessary usability engineering expectations, for example, that manufacturer is likely to conform with IEC 62366-1 Medical Devices— Part 1: Application of Usability Engineering to Medical Devices and should feel confident, if they have done their job correctly, that they have met the baseline for this aspect of medical device design. For medical device cybersecurity, the closest is the UL 2900 series, which, while not under the umbrella of the more common ISO or IEC, has been recognized and referenced by the regulators of numerous major markets across the globe. One of the valuable aspects of UL 2900 is that it covers the majority of the important aspects of medical device security, including security risk management, quality system, testing, and controls.

UL 2900 is split into 2 parts, the more general UL 2900-1 and the healthcare-specific version, UL 2900-2-1 Outline of Investigation for Software Cybersecurity of Connectable Products, Part 2-1: Particular Requirements for Connectable Components of Healthcare Systems. Both standards should be used in coordination to achieve the full intended scope of expectations. UL 2900 is also the basis for the only currently available certification of medical device security, through the UL Cybersecurity Assurance Program (CAP).

Cybersecurity risk management is another subspecialty that is important to consider. Safety risk management is well-established in the medical device industry, based on the approach found in ISO 14971. Cybersecurity risk management follows the same basic process steps, with several notable differences. AAMI TIR 57 has been a well-received approach for outlining the unique expectations for cybersecurity risk management of medical devices. It recommends a parallel approach whereby safety and cybersecurity risk management processes are maintained in a parallel but connected process where all cybersecurity risks are analyzed and controlled in the cybersecurity risk management process but, when a cybersecurity risk is identified as a potential safety issue, it gets propagated and added as a new risk within the safety risk management process. One of the primary reasons for the duplication of such risks is that the assessment criteria vary between these two processes. Safety risk management is based on assessment criteria that measure occurrence rates (probability, likelihood) and severity. Cybersecurity risk management, by contrast, typically assesses risk by measuring exploitability and impact. This is mainly because cybersecurity threats are not statistically predictable and it is more appropriate to use ease of exploitation of a given vulnerability as a measure (i.e., how easy would it be for an adversary to exploit this vulnerability should he/she decide to do so). Sources of security risk information are also unique, requiring the monitoring for new vulnerabilities and threats as part of the ongoing postmarket risk management process.

The National Institute for Standards and Technology (NIST) is a U.S. organization founded in 1901 and is one of the nation's oldest physical science laboratories. One of the more valuable aspects for the security community is the deep breadth of special publications released and managed by NIST. These publications are divided into series and special publication series of interest include Special Publication Series 800: Computer Security and Series 1800: General Information. NIST also publishes a variety of cybersecurity case studies, including a series of case studies in cyber supply chain risk management. NIST can be accessed at www.nist.gov. A list of useful NIST Special Publications can be found in the tables below (Tables 2.2–2.5). One commonly used NIST document is NIST SP 800-53, which is an important resource document for meeting government security expectations when selling medical devices to government organizations such as the U.S. Department of Veterans Administration.

The standards environment for medical device security often revolves around four basic sources: (1) health care specific standards, (2) general security standards, (3) industrial control security standards, and (4) medical device software standards. In addition, there are numerous technical documents that serve as excellent supporting material. For example, because many medical devices are based on embedded software, industrial control standards have proven helpful. The following tables provide a selection of these standards and technical reference documents to support medical device manufacturers in their product security program goals. These lists should not be considered comprehensive. However, they represent a diverse set of resources from which to select the most applicable.

2.3.2.1 Additional NIST Special Publications

- NIST SP 800-30 Risk Management Guide for Information Technology Systems;
- NIST SP 800-160 Systems Security Engineering;
- NIST SP 800-64 Security Considerations in the System Development Lifecycle;
- NIST SP 800-61 Computer Security Incident Handling Guide;

Table 2.2

Health Care-Specific Standards and Technical Guidance

Document Title	Description
UL 2900-1 Software Cybersecurity for Network-Connectable Products, Part 1: General Requirements	Applies to network-connectable products that shall be evaluated and tested for vulnerabilities, software weaknesses and malware: (i) developer risk management process requirements; (ii) methods to test vulnerabilities, software weaknesses and malware; and (iii) security risk control requirements.
UL 2900-2-1 Software Cybersecurity for Network-Connectable Products, Part 2-1: Particular Requirements for Network Connectable Components of Healthcare and Wellness Systems	Requirements for network connectable components of health care and wellness systems. This is a security evaluation standard that applies to medical devices, accessories to medical devices, and medical device data systems.
AAMI TIR 57 Principles for medical device security—risk management	Methods to perform information security risk management for medical device within the context of ISO 14971. Incorporates the view of risk management from IEC 80001-1.
AAMI TIR 97 Principles for medical device security —Postmarket risk management for device manufacturers	Focuses on the specific challenges of managing medical device security in the postmarket lifecycle phase, including vulnerability monitoring, designing with postmarket considerations, and disclosure programs.
IEC 80001-2-2 Application of risk management for IT—networks incorporating medical devices—Part 2-2: Guidance for the disclosure and communication of medical device security needs, risks, and controls.	Includes the 19 security capabilities used in the manufacturers disclosure statement for medical device security (MDS^2) form, including authorization, authentication, and emergency access.
IEC 80001-2-8 Application of risk management for IT-networks incorporating medical devices	Maps the 19 security capabilities in IEC 80001-2-2 to security controls in other existing standards.
IEC 80001-2-9 Application of risk management for IT-networks incorporating medical devices	Establishes a security case framework and provides guidance to health care delivery organizations (HDO) and medical device manufacturers (MDM) for identifying, developing, interpreting, updating and maintaining security cases for networked medical devices.
ISO 27799 Health informatics— Information security management in health using ISO/IEC 27002	Explains organizational information security standards and information security management practices including the selection, implementation and management of controls taking into consideration the organization's information security risk environment(s).
NIST SP 1800-24, Securing Picture Archiving and Communication System (PACS)	Intended to help HDOs implement current cybersecurity standards and best practices to reduce their cybersecurity risk while maintaining the performance and usability of PACS.
NIST SP 1800-8 Securing Wireless Infusion Pumps in Healthcare Delivery Organizations	Assists HDOs to implement current cybersecurity standards and best practices to reduce their cybersecurity risk while maintaining the performance and usability of wireless infusion pumps.
NIST SP 1800-1 Securing Electronic Health Records on Mobile Devices	Characteristics and capabilities that an organization's security experts can use to identify similar standards-based products that can be integrated with a healthcare provider's existing tools and infrastructure.

Table 2.3
General Security Standards and Technical Guidance

Document Title	Description
ISO/IEC 29147 Information technology—Security techniques—Vulnerability disclosure	Details the methods a vendor should use for the disclosure of potential vulnerabilities in products and online services.
ISO/IEC 30111 Information technology—security techniques—vulnerability handling process	Explains how to process and resolve potential vulnerability information in a product or online service.
ISO/IEC 27000 Information technology—Security techniques—Information security management systems—Overview and vocabulary	Provides the overview of information security management systems (ISMS). It also provides terms and definitions commonly used in the ISMS family of standards.
ISO/IEC 27001 Information technology—Security techniques—Information security management systems—Requirements	Well-known standard for providing requirements for information security management systems. Organizations are often certified to this standard.
ISO/IEC 27002 Information technology—Security techniques—Code of practice for information security controls	Guidelines for organizational information security standards and practices including the selection, implementation and management of controls taking into consideration the organization's information security risk environment.
NIST SP 800-53 Security and Privacy Controls for Federal Information Systems and Organizations	Recommends security controls for federal information systems and organizations and documents security controls for federal information systems

- NIST SP 800-122 Guide to Protecting the Confidentiality of Personally Identifiable Information;
- NIST SP 800-48 Guide to Securing Legacy IEEEE 802.11 Wireless Networks;
- NIST SP 800-41 Guidelines on Firewalls and Firewall Policy;
- NIST SP 800-167 Guide to Application Whitelisting;
- NIST SP 800-185 SHA-3 Derived Functions: cSHAKE, KMAC, TupleHash, and ParallelHash;
- NIST SP 800-133 Recommendation for Cryptographic Key Generation;
- NIST SP 800-144 Guidelines on Security and Privacy in Public Cloud Computing;
- NIST SP 800-121 Guide to Bluetooth Security;
- NIST SP 800-153 Guidelines for Securing Wireless Local Area Networks (WLANs);
- NIST SP 800-179 Guide to Securing Apple OS X 10.10 Systems for IT Professionals: A NIST Security Configuration Checklist;

Table 2.4
Industrial Control Standards and Technical Guidance

Document Title	Description
ISO/IEC 15408 (series) Evaluation criteria for IT security	Common criteria. Establishes general concepts and principles of IT security evaluation, models for evaluation of security properties of IT products.
IEC TS 62443-1-1 Edition 1.0 2009-07 - Industrial communication networks - Network and system security - Part 1-1: Terminology, concepts and models.	Introduces the concepts, terminology, and models used throughout the series.
IEC 62443-2-1 Edition 1.0 2010-11 - Industrial communication networks - Network and system security - Part 2-1: Establishing an industrial automation and control system security program	Describes what is required to define and implement an effective IACS cybersecurity management system.
IEC TR 62443-3-1 Edition 1.0 2009-07 - Industrial communication networks - Network and system security - Part 3-1: Security technologies for industrial automation and control systems.	Technical report describes the application of various security technologies.
ISO IEC 62443-3-3 Security for industrial automation and control systems Part 3-3: System security requirements and security levels	Describes the foundational system security requirements and security assurance levels.
ISO IEC 62443-4-1 Security for industrial automation and control systems Part 4-1: Product security development lifecycle requirements	Describes the derived requirements that are applicable to the development of products. The principal audience include suppliers of control systems solutions.

Table 2.5
Software-Specific Standards and Technical Guidance

Document Title	Description
IEC 62304 Medical device software—Software life cycle processes	Defines the lifecycle requirements for medical device software: the set of processes, activities, and tasks including development, maintenance, configuration management, and problem resolution.
IEC 82304-1 Health software—Part 1: general requirements for product safety	Covers entire lifecycle including design, development, validation, installation, maintenance, and disposal of health software products. Covers safety and security of health software products designed to operate on general computing platforms and intended to be placed on the market without dedicated hardware.

- NIST SP 800-183 Networks of Things;

- NIST SP 800-184 Guide for Cybersecurity Event Recovery;

- NIST SP 800-161 Supply Chain Risk Management Practices for Federal Information Systems and Organizations;

- NIST SP 800-16 Information Technology Security Training Requirements: a Role- and Performance-Based Model;
- NIST SP 800-40 Guide to Enterprise Patch Management Technologies;
- NIST SP 800-115 Technical Guide to Information Security Testing and Assessment;
- NIST SP 800-70 National Checklist Program for IT Products: Guidelines for Checklist Users and Developers;
- NIST IR 8053 De-identification of personally identifiable information;
- NIST SP 1800-11 Data Integrity: Recovering from Ransomware and Other Destructive Events;
- NIST SP 1800-4 Mobile Device Security: Cloud and Hybrid Builds.

Successful and effective product security programs for medical devices include considerations of both the regulatory and standards environment. Obviously, passing through the regulatory hurdles is a necessary activity to gaining market entry into each region, but it can also reflect expectations from the customer base as well. Hospitals are also experiencing pressure to address cybersecurity from an operational as well as a privacy perspective. Therefore, security considerations are often being integrated into purchasing decisions, vendor approval processes and formalized into contracts. The ability to demonstrate that a vendor and its products meet a standard bar of acceptability can be necessary before a hospital will approve the purchase of a connected medical device, particularly if that medical device connects to the hospital network or other shared systems. The ability to demonstrate conformity to standards and regulatory expectations can be key to proving that a medical device has been designed and is managed in a way that security is sufficiently addressed. Increasingly, hospitals are appreciating that their networks are only as secure as their weakest link, so they have been more selective in their review and approval of medical devices that cannot provide evidence of a robust security profile. As the industry continues to mature, all stakeholders, including regulators, hospitals, and medical device manufacturers, are likely to raise the bar on security. This proactive approach is necessary to stay ahead of the threat actors who are also evolving and maturing.

2.4 Security and Lifecycle Management: High-Level Overview

A lifecycle-based approach to medical device cybersecurity is actually the combination of four distinct lifecycles and the interaction between them (see Figure 2.2). Besides managing the complexities and intricacies of each of the individual cycles, a challenge for most MDMs is to understand and interface with the components that are not under their direct control, yet are required to function in harmony to provide the desired outcome. In other words, in addition to the internal processes it also becomes an issue of the interfaces to other stakeholders (i.e., the flow of information and coordination with their respective processes). This may require contractual agreements, process definitions, or organizational memberships (as for example in an information sharing and analysis organization (ISAO)). But most importantly, it requires cultural change that embraces a willingness to share and communicate.

As with any model, the lifecycles depicted in Figure 2.2 are an abstraction that illustrate the most general case. It cannot reflect all the details and variations that are required in the real world. However, it does provide a usable reference architecture and framework that can be followed and can be used for explanatory purposes, as it is in this book.

The four key lifecycles we need to be aware of and manage, discussed in further detail through the remainder of this book, are:

1. Manufacturer development lifecycle: From a regulatory perspective, the development lifecycle encompasses what is typically referred to as premarket activities (i.e., the steps required prior to a product being approved for and introduced to the market). Starting with the inception of a new product or version, to design steps, to manufacturing transfer, postmarket management, and eventually end-of-life, it is the manufacturer's leadership that is ultimately responsible for enabling an organization to successfully deliver secure products to the market.

   The typical steps for a security-enabled development lifecycle are not necessarily different than in traditional software engineering, however, cybersecurity requires some unique considerations—for example, the need to make certain security review and testing activities a continual aspect of engineering activities, rather than a distinct step at the end. Furthermore, cybersecurity can be very dynamic and manufacturers are required to respond to newly discovered vulnerabilities, evolving threats, or customer-reported incidents no matter whether a product is already in the market or if it is still in development.

   In today's complex and global business world, not all these phases may be conducted by the manufacturer. To complement internal talent, capability, and capacity, certain development, testing, or manufacturing tasks may be conducted under an outsourcing arrangement. But even though the manufacturer may not conduct all of these steps, it still needs to control them because it is the manufacturer and distributor of the finished product who ultimately has regulatory and legal responsibility.

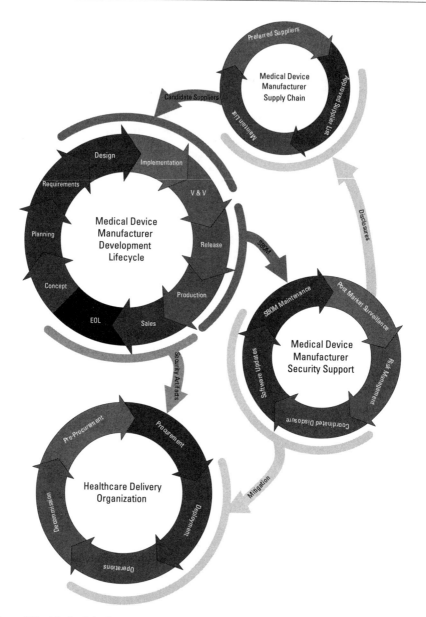

Figure 2.2   Medical device cybersecurity lifecycle components. (©H-ISAC. Used with permission.)

We usually consider these activities as circular, rather than linear processes with a beginning and end. The conception and development of the next version or next generation of a product will typically be already well on its way when the current version reaches the end of its commercial life. Furthermore, lessons learned from the previous product or version should be incorporated in the planning and requirements development for the next one, as discussed in more detail in Section 2.8 of this chapter.

2. Manufacturer maintenance lifecycle: In the postmarket phase the manufacturer is expected to undertake a set of activities that, collectively, are to assure the maintenance of a device's security posture and to enable response to cybersecurity signals (to use FDA terminology). The key element of these activities is the SBOM because it defines the device's composition at the level of its individual component versions, no matter if these are third party or internally developed components.

   As part of the manufacturer's postmarket surveillance, any security risks identified (i.e., new threats or vulnerabilities) require appropriate and timely assessment and response. The information could be obtained from or provided by customers, service personnel, regulators, security researchers, suppliers, or device user and could be based on observation of the threat landscape, analysis of reported incidents, disclosures of newly discovered vulnerabilities, formal complaints or regulatory action, returned products, or service records.

   Should the assessment of the respective cybersecurity risk warrant action, a coordinated disclosure to the customer base will provide the information required for the customer to understand the impact and decide on appropriate action. Furthermore, it may require release and distribution of a software update (patch), which then in turn will be reflected in an updated SBOM for the device.

   Note that these are just generalized process steps that may vary based on device design and risk specifics. For example, a design change may be caused by a hardware vulnerability or may lead to hardware changes or updates to hardware-residing microcode. Although commonly a new risk is based on a newly discovered vulnerability or threat, there could also be the case that a change in the device's intended use requires an updated risk assessment because a risk previously defined as acceptable may no longer be so in the new use case.

3. Supply chain lifecycle: Modern medical devices are complex and so is their technology ecosystem. Today's medical device manufacturers commonly contract out part of their design or utilize commercial off-the-shelf (COTS) or open-source software.

   It is essential that these software components, and partners supplying them, are properly managed to assure that any third-party component is managed to the same standard as the in-house developed software. This does not require that the supplier itself complies with FDA regulations and has supporting processes in place—although some do. However, the supplied software will need to be integrated, validated, and documented under the device manufacturer's quality system and applicable regulations.

Common components that device manufacturers incorporate from external suppliers are operating systems, drivers, signal and image processing algorithms, and interface software. However, it may also include highly specialized components, custom solutions, and software addressing infrastructure needs, for example, complementary security software like antimalware or encryption.

4. HDO lifecyle: This covers the typical activities required on the HDO side from security-inclusive preprocurement planning, execution of the procurement process, onboarding and deployment of devices, operation and maintenance, and ultimately the end of useful clinical life and proper decommissioning.

The security objectives and requirements for the clinical lifecycle are to reduce (to an acceptable level) the risks associated with patient safety, information privacy, care delivery disruption, and business exposure.

2.4.1   Coordination between the Four Lifecycles

These four lifecycles, although managed independently by different stakeholders, have a relationship at certain touchpoints that serve specific functions to assure proper management of security requirements and, ultimately, delivery of a proactively secured device:

- *Supplier management:* The MDM's management of its supply chain is a continual, ongoing process as it requires careful selection of preferred suppliers, enforcement and maintenance of an approved supplier list (ASL), and management of the supplier and supplied software. The approved supplier candidates include both custom and off-the shelf software manufacturers and will be phased into the medical device product during design, implementation, and verification and validation (V&V).

- *Sales process:* As part of the sales process and product delivery, the MDM typically provides a set of security artifacts to the HDO, including but not limited to:

  - Hazard analysis, mitigations, and design considerations (including list of cybersecurity risks considered and cybersecurity controls applied);

  - Traceability matrix that links cybersecurity controls to the respective risks;

  - A plan for postmarket management of software updates and patches;

- Documentation of software integrity controls (i.e., assurance that the software remains free of compromise);
  - Instructions and specifications for cybersecurity controls (e.g., anti-malware software or firewall management), as well as for secure deployment, handling, and integration.

- These artifacts include, but are not limited to, documentation of: security risk assessment, device software composition (SBOM), service security properties (MDS^2), as well as assurance of software security (e.g., evidence of testing or certification). Agreement on the supporting documentation may be formalized in form of a contract between the parties and should enable the HDO to securely deploy and maintain the device during its useful clinical life.

- *MDM maintenance:* This is the most connected of the lifecycles and requires coordination between the MDM, any third-party supplier to the MDM (via disclosure of any issues that may be discovered as part of postmarket surveillance), and the HDO (via coordinated disclosure and mitigation of any newly discovered vulnerabilities or threats). The reference point for these activities is the device's SBOM, which enables surveillance relative to the SBOM components as well as management, disclosure, and mitigation of any issues uncovered.

This section reviewed the four main lifecycles that govern the steps and processes that enable us to manage the regulatory, engineering, operational, and maintenance processes. This high-level overview was focused on developing a general understanding of the lifecycle-based approach and how they apply to and connect the individual stakeholders. This will be discussed in much more detail in the remainder of this book.

2.5 Regular Review of Security Maturity

Whether an MDM is just setting out on the journey towards becoming a cyber-capable organization or already has taken steps in the right direction, understanding one's current security maturity, defining the desired future state, understanding external drivers and changes, and measuring progress along the path are of critical importance. For that purpose, security maturity models can be applied, enabling an organization to assess, plan, and adjust.

In general, a maturity model is a tool that allows an organization to understand its effectiveness at achieving a particular goal, in this case cybersecurity capabilities (engineering, manufacturing, and communication processes) and outcomes (the delivered end product). A good security maturity model gives an organization's leadership a reliable measure on its current capabilities, desired future state, and tangible actions to take to move up the maturity curve. It aligns business goals with day-to-day operations and enables measuring progress.

Any organization, its employees, and applied practices require review and adjustment on a regular basis. Processes may be found to have shortcomings that can be improved on, or the environmental conditions may change due to new regulatory requirements, changes in market needs, or newly discovered cyber risks. It is essential that an organization stays vigilant by observing these changes, identifying gaps, and developing and progressing along a path forward.

Assessing security maturity requires a regular "so, how are we doing?" exercise with input provided from internal and external stakeholders. It also requires that an organization has identified the data sources that are consulted for this purpose, implements regular assessment milestones, and defines the participants that manage the maturity review process or provide input to it.

Improving an organization's security capabilities and maturity is both a tactical step-by-step exercise and a long-term strategy (where do we want to be in x years?). Did the organization embrace security practices and demonstrate security culture, or is it an overlay that is forced onto its processes and employees?

Typical events that provide grounding and instigation for a maturity review are the major handoff steps during the lifecycle, as well as both key events and predetermined review intervals taking place after product has been released and has some market experience under its belt. What worked, where did we fail? Can we improve efficiency, reduce errors, or improve the product? These are typical questions that should be asked.

A security maturity review should be a complete, end-to-end undertaking. Is the finished product sufficiently secure and did it meet regulatory and customer requirements? Does it deploy and operate as intended and is it capable to withstand the security challenges of its normal operating environment?

Maturity reviews should always include both an assessment of the success of the device's design (does it meet the defined security requirements?) as well as a review of the quality of the processes that were applied (did their use result in a sufficiently secure product?). Furthermore, prior to launching the next development cycle (new product or update) one should assess if any of the environmental variables have changed (target markets, use cases, regulations, threat landscape, etc.), since they likely have. This then should result in an update of processes, training and skill set, design requirements, and technology used.

Security maturity reviews are commonplace in traditional IT security. Models exist that can be applied to or adopted for assessing security maturity of a security engineering organization and the applied lifecycle processes. Some examples for security maturity models are: The Systems Security Engineering Capability Maturity Model (SSE-CMM) [9] (security engineering and software design), NIST's Program Review for Information Security Assistance

(PRISMA) (federal, critical infrastructure) [12], Department of Homeland Security's Cybersecurity Capability Maturity Model White Paper [11] (process and capability), or European Union Agency for Cybersecurity: ENISA CSIRT Maturity Assessment Model [12] (organization, human, tools, processes).

There is a saying in cybersecurity that "you can't protect what you can't see." The analogous statement for security maturity would be: "you can't improve what you don't assess." Security maturity models fill this gap.

References

[1] Morgan, C., "Medical Device Cybersecurity Regulatory Publications," Apraciti, Nov. 26, 2019, https://www.apraciti.com/blog/2019/11/25/global-regulatory-authority-publications-on-medical-device-cybersecurity.

[2] U.S. Food and Drug Administration (FDA), "Cybersecurity," Jan. 23, 2020. https://www.fda.gov/medical-devices/digital-health/cybersecurity.

[3] U.S. Department of Homeland Security, "ICS Medical Advisory (ICSMA-19-274-01)," ICS-CERT Advisories, Jan. 7, 2020, https://www.us-cert.gov/ics/advisories/icsma-19-274-01.

[4] Royce, E., "H.R.5793 - Cyber Supply Chain Management and Transparency Act of 2014," December 4, 2014, https://www.congress.gov/bill/113th-congress/house-bill/5793.

[5] Bellissimo, A., J. Burgess, and K. Fu, "Secure Software Updates: Disappointments and New Challenges," *Proceedings of the USENIX Summit on Hot Topics in Security*, Vancouver BC, Canada, Jul. 31–Aug. 4, 2006, https://www.usenix.org/legacy/event/hotsec06/tech/full_papers/bellissimo/bellissimo.pdf.

[6] Burleson, W., S. Clark, B. Ransford, and K. Fu, "Design Challenges for Secure Implantable Medical Devices," *Proceedings of the 49th Annual Design Automation Conference*, San Francisco, CA, June 2012, pp. 12–17, https://spqr.eecs.umich.edu/papers/49SS2-3_burleson.pdf.

[7] Fu, K., "Trustworthy Medical Device Software," *Public Health Effectiveness of the FDA 510 (k) Clearance Process: Measuring Postmarket Performance and Other Select Topics: Workshop Report*, Washington, DC: The National Academies Press, 2011, p. 102, https://doi.org/10.17226/13020.

[8] Government Accountability Office (GAO), "FDA Should Expand Its Consideration of Information Security for Certain Types of Devices," Aug. 31, 2012, https://www.gao.gov/products/GAO-12-816.

[9] Systems Security Engineering Capability Maturity Model (SSE-CMM), https://www.sse-cmm.org/model.htm.

[10] National Institute of Standards and Technology (NIST), "Program Review for Information Security Assistance Project," Mar. 29, 2018, https://csrc.nist.gov/Projects/Program-Review-for-Information-Security-Assistance.

[11] Department of Homeland Security, "Cybersecurity Capability Maturity Model," Aug. 4, 2014, https://niccs.us-cert.gov/sites/default/files/Capability%20Maturity%20Model%20White%20Paper.pdf.

[12]   European Union Agency for Cybersecurity (ENISA), "Challenges for National CSIRTs in Europe in 2016: Study on CSIRT Maturity," Apr. 30, 2019. https://www.enisa.europa.eu/publications/study-on-csirt-maturity.

3

Supply Chain Management

3.1 Upstream Supply Chain Management

The U.S. Code of Federal Regulations part 820 section 50 (CFR 820.50) states that "Each manufacturer shall establish and maintain procedures to ensure that all purchased or otherwise received product and services conform to specified requirements." [1]

For the last several years, this has been an area of focus for both the FDA and international regulatory bodies. Activities expected under this broad conceptual umbrella include:

- Formal identification, inspection, and documentation of quality-critical attributes of purchased or received product and services;
- Establishing and enforcing processes to control and manage vendors, including collecting consistent and reliable data on all suppliers.

More recently, security has been added to the list of supplier attributes that manufacturers are expected to monitor (Figure 3.1). This has become an attribute of heightened concern and scrutiny because when it comes to security, the supply chain is just that, a chain: only as strong as its weakest link. Attackers know this and are more likely to direct attacks against smaller, less-resourced companies in the supply chain, hoping to leverage an easier exploit opportunity against the larger, better-resourced, more attractive target than the smaller company services.

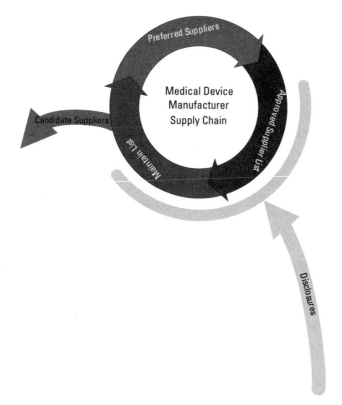

Figure 3.1 Supply chain lifecycle.

One of the authors was recently made aware of a prime example of this type of malicious activity when one of our clients was acquired by a Fortune 500 company. On the day the acquisition was announced, our client experienced a thousandfold increase in attempts to breach their security. The announcement had alerted attackers to the fact that our client was part of the supply chain of a prime target.

Manufacturers need to take steps to protect their devices from the following supply chain attacks:

- Computing platforms being preinfected with malware;
- Malware insertions into device software or electronic components;
- Vulnerabilities introduced into the software;
- Counterfeit components and subassemblies.

Upstream supply chain management is a difficult and expensive endeavor and is divided into two main areas: electronics and third-party software components.

3.1.1 Counterfeit Electronic Components

Several years ago, components started appearing on the market that was branded as the correct component in the correct type of packaging. However, the internal die was either completely different, or the component was relabeled and up-rated (e.g., labeling a standard component as MIL-Spec), or, in some cases, just a film of copper.

These components where usually repurposed components scavenged from e-waste recycling dumps, such as those found in China's Guangdong Province. These scavenged components are stripped from a PCBA by noncontrolled heating methods, washed in whatever water is available, and handled without any ESD protection. Components are then sanded or subjected to an acid bath, recoated, and remarked with the counterfeit component's branding. This process is typically referred to as blacktopping.

Visual inspection of incoming blacktopped components proved minimally effective. Inadequate quality of the rebranding or misalignment of the leads (for leaded components) sometimes resulted in detection during receiving. However, many blacktopped components made it through initial inspection and onto the final assembly floor, where they were generally detected by testing during manufacturing.

Because blacktopped components generally cannot perform to specification (or at all), they have rarely made it past the manufacturing floor and therefore rarely posed problems for end-users. However, they have been responsible for expensive and sometimes embarrassing delays in manufacturing.

3.1.1.1 Newer Counterfeits

The second form of counterfeit component appeared in the market more recently, one which not only appears to be correct but performs the correct component's intended operation, at least to some degree. This proved to be a much harder counterfeit to detect. For example, it is believed that many FTDI FT232RL USB-to-serial components were counterfeits containing a ROMed microcontroller that was emulating the original FT232 functionality [2]. Of course, this emulation was not completely accurate to the original FT232 component, which could cause impacts on many of the devices in which it was placed.

The FTDI component is just one example of many [3], including the ATMega328 [4] and MicroSD cards. Probably most surprising is the counterfeiting of extremely low-cost (but high-volume) components such as diodes, capacitors, and resistors.

Usually, components are counterfeited when the cost of the original, genuine components has increased. This can be due to availability shortages, natural and other disasters impacting production or transportation, or fluctuations in the costs of raw materials. The bottom line is, no component can be assumed safe from counterfeiting.

Besides visual inspection of the components, there are other means to confirm the validity of the component. These include both destructive and nondestructive tests.

An example of destructive testing (applied to a representative sample from each component lot) is decapping the device and visually comparing it with the target using a microscope to ensure the die surface is consistent with the specification.

X-ray stations allow for the verification of internal lead frames, bonding wires, and die construction. This can be performed by third party testing houses or on-premise X-ray machines, such as the Glenbrook Technologies JewelBox product line. Some types of components may be destroyed by X-ray testing, while other types may not.

A nondestructive test method is to analyze multiple unique electrical characteristics of a component. This is the approach taken by Battelle's Barricade test station. This method is nondestructive, low-cost per component, and can be applied to all components (not just a lot sample), and can be configured for high throughput rates.

3.1.1.2   Inspection Triggers

The following are some red flags that should trigger inspections or, depending on when in the process the red flag is encountered, prompt further action to verify that the component is genuine and trustworthy:

- Item comes from a source other than the OEM or authorized source;
- Item comes from an unknown source;
- Item comes from a source in a suspect geopolitical location (e.g., component stores in China);
- Item is priced too low/significantly different from historical pricing;
- Scarce components are suddenly available in large quantities;
- Chain of ownership unverifiable;
- No certificate of conformance;
- Aberrations of size/shape/color/finish/materials;
- Evidence of rework/repair/refinishing/resurfacing of the component;
- Component leads are deformed or exhibit alignment issues;
- Component labeling/branding issues:
  - Does not match previous items of the same type;
  - Incomplete orders;

- Incorrect size of packaging;
- Lot number/date code issues;
- Shoddy package/construction issues;
- Unusual packaging medium (reel vs. tray, etc.).

For advanced needs for electronic component inspection, including computer chips, several research institutions have developed advanced reverse engineering capabilities, for example, the REFINE lab at the University of Connecticut and many other commercial service providers [5].

Management of your Approved Supplier List (see Section 3.2) needs to include results from counterfeit detection events. In general, avoid gray market suppliers of electronic components, contractually hold contract manufacturers to sourcing components from reliable sources, and don't migrate to new sources without prior approval.

3.1.1.3 Subassemblies

Subassemblies pose a much harder threat to mitigate than individual components since you are now exposed to a tree of potential counterfeit sources, most or all of which may be hidden from your knowledge.

Compliance with key security requirements has to be contractually specified between yourself and the vendor of the subassembly. The vendor likewise needs to contractually enforce similar security requirements with his vendors, and so on down the supply chain, until the component level is reached. This chain of contractually clarified trust must also be specified in the contract between the MDM and the subassembly vendor.

The mechanism used to convey this trust is a Certificate of Conformance (CoC) for each shipment. The CoC is a formal document signed by the vendor and given to the MDM. It affirms that all purchase order requirements have been met and these requirements should include not only specific information about the parts, such as quantity and lot and date codes, but also sourcing information about the vendor's vendor, including OEMs and all distributors contributing to the build-out of the subassembly.

3.1.1.4 Disposal of Counterfeit Components/Subassemblies

Counterfeiting invariably involves fraud, often from an upstream supplier that can be several tiers removed from the device manufacturer itself. Therefore, coordination with appropriate officials must take place prior to disposing of counterfeit parts. If the suspect items are contained within your facility, the MDM should:

- Maintain counterfeit parts or materials in quarantine, clearly identified as counterfeit materials;

- Never return counterfeit materials to the vendor under any circumstances, as they could be reintroduced into the supply chain to be sold again to another victim.

Legal authorities should be contacted to initiate an investigation into the counterfeiting activity. If such an investigation is performed, then the counterfeit materials may be required as evidence.

3.1.2 Third-Party Software Components

The other primary area of upstream supply chain management concerns third-party software components (TPSC). Whether referred to as SOUP (an acronym variously defined as software of unknown provenance/pedigree/parentage) or TPSC, we are talking about open- or closed-source libraries, frameworks, hardware drivers, runtime engines (e.g., .NET Framework), and operating systems utilized in the medical device system. In this book we will use the more accurate label TPSC.

TPSCs are software components integrated into the actual finished product. These are not to be confused with:

- The toolchain being used to compile the code;
- The operating system on the workstation used to build the code;
- The programming editor/IDE;
- The continuous integration tools;
- The bug tracking tool;
- The version control tool.

Each of these merits its own security considerations and should be controlled under the purview of the MDM's infrastructure and IT security framework. However, none of them fall under the category of TPSC.

The amount of TPSC utilized in a product is largely dependent upon the nature of the embedded medical device. For example, a small coin cell battery-operated device containing an 8-bit, 16-bit, or 32-bit microcontroller will usually incorporate much fewer TPSCs (typically fewer than 10) compared to an embedded PC or a smartphone app, where hundreds of TPSCs are typical.

TPSC-based attacks have been ramping up in frequency of occurrence, especially since 2018. Twelve popular open-source libraries were compromised through a variety of exploits [6], including in several cases convincing the old moderator to turn over control of the TPSC to a new moderator [7], who then promptly injected exploit code into the repository. However, these attacks are not just upon open-source libraries, but also closed-sourced programs as well [8].

Interestingly, those referenced cases are linked. The CCleaner malware is thought to have given the attacker access into Asus, which then allowed them to steal the signing credentials for the Asus firmware update server, making this an example of a chained attack upon the supply chain!

The rapidly growing threat of supply chain attacks has even prompted DHS to get involved [9], establishing a Supply Chain Risk Management Task Force in November 2018.

Practically speaking, MDMs can do very little to avoid TPSC-related problems beyond layering the system mitigations in a defense-in-depth approach and leveraging SBOMs to monitor for disclosed vulnerabilities resulting from a compromised TPSC.

At the time of regulatory premarket submission, a required element of the submission package is a review of the TPSCs utilized in the system and a justification for the inclusion of any TPSC with a known vulnerability. This justification should take the form of a patient risk-benefit analysis. Please refer to Chapter 8 for more information about TPSC disclosure and justification.

The review of disclosed vulnerabilities should also include any subsequent versions of the TPSC that have been released since the older version that is being used in the medical device system. This information about all utilized TPSCs will also be codified in a machine-readable format as an SBOM. Please refer to Chapter 8, SBOMs for more information.

Supply chain management can't be achieved with just a technical solution; most controls are people-powered. Similarly, this isn't just an engineering department process, rather it is a process that involves many groups within the company performing various activities to secure the supply chain.

Supply chain management is difficult and costly—but not as costly as failing to secure the supply chain. It should not be taken lightly, nor assumed that some other group in your organization is performing it. The processes need to be documented, performed, and tested before any confidence in the supply chain or your products can be achieved.

3.2 Security Criteria for Approved Supplier Lists

In controlled research and development (R&D) and controlled production, all suppliers are vetted against established criteria for inclusion on an Approved Supplier List (ASL). Every item on a proposed BOM is sourced from a supplier in good standing on the ASL, swapped out for one that is, or designed out of the system.

Typical ASL vetting criteria includes items such as:

• Quality and maturity of the component production line;

• Responsiveness and effectiveness of customer service;

- Contract and confidentiality terms the supplier considers acceptable or unacceptable;
- Disclosure of details about the supplier's facilities, operations, business culture, quality system, and outsourcing;
- Control of subsuppliers and other business partners;
- Disaster recovery and incident response plans and readiness;
- Responses from supplier-provided references.

MDMs wishing to engage in the secure development lifecycle will need to expand their existing ASL practices to include cybersecurity if they have not already done so. This expansion will impact their existing ASL auditing procedures, trigger reevaluation (and possibly reclassification) of active suppliers already in good standing on the list, and could ultimately have impacts on the BOMs of projects currently in development. Allowances should be made for the potential ripple effect, keeping in mind that these changes are necessary to achieve proper supply chain management.

Cybersecurity considerations for ASLs include the following:

- Expansion of the list to encompass not just suppliers on the hardware BOM, but the software BOM (SBOM) as well. Traditional ASL considerations will need to be evaluated and perhaps modified or excluded, based on their applicability to TPSC suppliers.
- Each supplier, whether of hardware and software, is considered untrusted when added to the ASL. Only sustained compliance and auditing can raise the trust level of a supplier.
- Auditing should be performed on the entire supply chain, not just the one immediately upstream of the MDM.
- When audits are performed, they should include a focus on security (both physical security and cybersecurity).
- Each of the suppliers in the chain must have standardized security processes, and be able to provide confirmation that these processes are being implemented.
- Each of the suppliers should perform regular Red Team tests of their organization.
- Red Teaming should stress:
  - Physical: including access control, temporary access, warehouses, offices, trash, server rooms, wiring closets, elevators;
  - Cyber: including routers, wi-fi, networks, IoT devices, printers, public and private access;

- People: staff, contractors, business partners, vendors, social engineering, coercion, bribery, data authorization levels.
- Open-source utilization at each vendor should be evaluated for license usage and adherence to each specific license type's constraints and requirements.
- Open-source should not be inherently trusted by any supplier in the chain, including the MDM.
- Attestation of the origin of the open source being utilized should be confirmed (i.e., not a forked variant of the original intended open source package);
- Changes in ownership, authorship, or moderation of a component being provided via the supply chain should reset that supplier to a non-trusted status until trust can be reaffirmed.
- Supplier rapid response to detected and/or disclosed vulnerabilities increases the preference for continued engagements with the vendor.
- Vendor proactiveness in notifying the MDM of all cybersecurity events, including breaches and vulnerabilities.
- Audits should physically confirm the disposition of any media containing the intellectual property of the vendor or the MDM.

It should be noted that although MDMs should prioritize the ASL according to inflexible must-meet criteria, there may be cases where the MDM must work with a supplier of a unique component that has no viable substitute, and the supplier is unable or unwilling to meet approval according to the MDM's established ASL procedure. In such cases, ASL procedure can allow a secondary, less-preferred approval pathway which waives the standard requirement(s) in favor of additional compensating controls and safety measures. These might include additional audits, inspection, contractual mitigation, source code escrow, and so forth.

3.3 Downstream Supply Chain Management

A downstream supply chain is typically referring to the consumers of your products, either HDOs, other health care providers, or home health care customers. But when cybersecurity is being considered, it also includes manufacturing and channel partner considerations. These considerations change, to some degree, depending on the geographic location of manufacturing and sales. MDMs often think of themselves at the end of the supply chain, however, that is rarely the case. From a device cybersecurity perspective, MDMs are somewhere in the middle of the chain.

Devices can be manufactured in or out of house. In either setting, integrity can be compromised and intellectual property can be exposed, but the MDM must use different means to assure the security of contract manufacturers than would normally be available in-house. Fulfillment houses can be involved with configuring and commissioning PCs intended for use in a medical device system, where malware can be installed intentionally or unintentionally by workers at the fulfillment house. Even the PC suppliers should not be trusted without verification [10] and should instead be ordered and supplied by the PC source as a barebones configuration. Failing that, the PCs should be completely wiped and reimaged before reaching the factory line. Finally, distribution [11] and warehousing can also be areas where the loss of integrity, IP exposure, and malware injection can take place. For an attacker, these are highly desirable targets, as they can infect many HDOs by only infecting one manufacturing or fulfillment location.

For the MDM, strong relationships and audits of this downstream supply chain are necessary to maintain any control of this process. Likewise, as devices are returned from the field for service, if any intentional corruption is discovered in the device it should be logged for further investigation and root-cause analysis of the downstream supply chain should be conducted to identify potential sources of the security breaches.

Also, the end-users (both HDOs and home health care patients) of the medical device should have functionality exposed to them to confirm the validity of the device integrity. This can include items such as log entries related to security issues, malware scanners, host intrusion detection, tamper evident packaging, cryptographic signatures, hashes, even an online presence that supplies nonrepudiation evidence to the end consumer. This enables the end consumer to have trust in the received device. However, this type of messaging or notification will widely depend on the role of the end-user, their technical capability, and their ability to respond to the message in a meaningful and constructive manner. Much of the postmarket relationship is now being standardized in contracts (see Chapter 4) between the MDMs and the HDOs. This means that the relationship between MDMs and HDOs will be tighter than ever, such that ensuring the integrity of the delivered medical device will be in the best interests of the MDM.

References

[1]   The U.S. Code of Federal Regulations, Title 21, Vol. 8, Part 820, No. 50, April 1, 2019, https://www.accessdata.fda.gov/scripts/cdrh/cfdocs/cfcfr/CFRSearch.cfm?fr=820.50.

[2]   Chirgwin, R., "FTDI Boss Hits Out at 'Chinese Criminal Gang' Pumping Knock-Off Chips," The Register, Feb. 10, 2016, https://www.theregister.co.uk/2016/02/10/ftdi_says_knockoff_chips_part_of_criminal_operation/.

"FTDI FT232RL: Real vs Fake," Zeptobars, Feb. 17, 2014, https://zeptobars.com/en/read/FTDI-FT232RL-real-vs-fake-supereal.

[3]   Villasenor, J., and M. Tehranipoor, "The Hidden Dangers of Chop-Shop Electronics," *IEEE Spectrum*, Sep. 20, 2013, https://spectrum.ieee.org/semiconductors/processors/the-hidden-dangers-of-chopshop-electronics.

[4]   Grady, E., "Revisiting the Counterfeit ATMega328s," SparkFun, May 17, 2010, https://www.sparkfun.com/news/364.

[5]   Smith Counterfeit Detection Lab: https://www.smithweb.com/counterfeit-detection-lab/.

Micross Counterfeit Mitigation Services: https://www.micross.com/electrical-test/counterfeit-mitigation/.

[6]   Goodin, D., "The Year-Long Rash of Supply Chain Attacks Against Open Source is Getting Worse," Ars Technica, Aug. 21, 2019, https://arstechnica.com/information-technology/2019/08/the-year-long-rash-of-supply-chain-attacks-against-open-source-is-getting-worse/.

[7]   Perekalin, A., "A Bad Link in the Cryptochain," *Kaspersky Daily*, Nov. 29, 2018, https://www.kaspersky.com/blog/copay-supply-chain-attack/24786/.

[8]   Goodin, D., "CCleaner Malware Outbreak is Much Worse than it First Appeared," Ars Technica, Sep. 21, 2017, https://arstechnica.com/information-technology/2017/09/ccleaner-malware-outbreak-is-much-worse-than-it-first-appeared/.

Zetter, K., "Hackers Hijacked ASUS Software Updates to Install Backdoors on Thousands of Computers," Vice, Mar. 25, 2019, https://www.vice.com/en_us/article/pan9wn/hackers-hijacked-asus-software-updates-to-install-backdoors-on-thousands-of-computers.

[9]   Department of Homeland Security, "DHS Announces ICT Supply Chain Risk Management Task Force Members," Nov. 15, 2018, https://www.dhs.gov/news/2018/11/15/dhs-announces-ict-supply-chain-risk-management-task-force-members.

[10] "Superfish Adware in Lenovo Consumer Laptops Violates SSL, Affects Companies via BYOD," Trend Micro, Feb. 20, 2015, https://www.trendmicro.com/vinfo/us/security/news/cybercrime-and-digital-threats/superfish-adware-in-lenovo-consumer-laptops-violates-ssl.

[11] Simpson, C., "The NSA Intercepts Laptops Purchased Online to Install Malware," The Atlantic, Dec. 29, 2013, https://www.theatlantic.com/technology/archive/2013/12/nsa-intercepts-laptops-purchased-online-install-malware/356548/

Masnick, M., "NSA Interception in Action? Tor Developer's Computer Gets Mysteriously Re-Routed to Virginia," Tech Dirt, Jan. 24, 2014, https://www.techdirt.com/articles/20140124/10564825981/nsa-interception-action-tor-developers-computer-gets-mysteriously-re-routed-to-virginia.shtml.

4

Medical Device Manufacturers' Development Cycle

4.1 Introduction

What is medical device cybersecurity? What are we trying to protect?

If you only read regulations and security standards, such as those provided by the U.S. FDA, Association for the Advancement of Medical Instrumentation (AAMI), Australia's Therapeutic Goods Administration (TGA), The International Medical Device Regulators Forum (IMDRF), UL, ISO/IEC, and so forth, you would believe that the process of securing a medical device is only about protecting the device's safety and efficacy for patients and end-users. But that isn't the whole story.

It may be true that many MDMs would not have adopted a secure development lifecycle for their medical device development without the mandate by the FDA and other global regulatory agencies. Oversight from these agencies is driving MDMs to create documentation artifacts to substantiate that their devices are sufficiently secure, with residual risk sufficiently low, for patients to use. The activities described in this chapter are the basis for these artifacts, which are submitted during premarket approval and scrutinized by the respective local regulatory agencies—for more details, see Chapter 8. These activities and artifacts are not difficult to perform and create, and they help MDMs ensure that they develop secure systems that risk neither end users' safety, nor the MDM's business interests.

MDMs are, of course, highly concerned with safety and efficacy, so that patients are not harmed through malicious action or failure of the essential performance of the medical device. However, there are many other areas and activities that are highly relevant to securing a medical device—and consequently, medical device security should not only be driven by regulatory consideration. Collectively, these other attributes of a secure medical device can be addressed as threats to the manufacturer's business model or, in contexts where we are clear about two uses of the term risk, we can use the more familiar phrase business risk.

Business risks cover areas such as:

- Loss of consumer confidence and subsequent market share;
- Cloning/counterfeiting of disposable accessories;
- Reenabling the use of consumed accessories that have reached their end of operational life;
- Extracting intellectual property, such as software and algorithms, from the medical device;
- Extracting security mitigation elements, such as cryptographic keys and shared secrets, to leverage in a larger subsequent attack.

In addition, there are business risks to the operator of the device (typically the HDO), that, should they manifest themselves, may lead the affected party to seek compensation or file lawsuits against the MDM:

- Exposure of PHI/PII and the associated legal fines for such a breach;
- Loss of business due to inability to deliver care;
- Loss of business due to reputational impact;
- Harm of patients or users due to a targeted or nontargeted attack on a device;
- Weaponizing the medical device to attack other entities, or to act as a pivot point into an intranet, such as a manufacturers' or HDO's intranet, or a patient's home network.

It should be noted that in some of these business risks, the patient can also be negatively affected—for example, by the potential poor performance of a cloned or reenabled accessory, or due to your device being utilized as a pivot point to attack other systems or the infrastructure on which those devices depends, thereby placing the entire organization and other patients at risk of harm [1].

In order to gain control across the entire range of risks, from regulatory and safety to business risks, medical device manufacturers (MDMs) are advised to adopt a secure development lifecycle of medical devices, as shown in Figure 4.1.

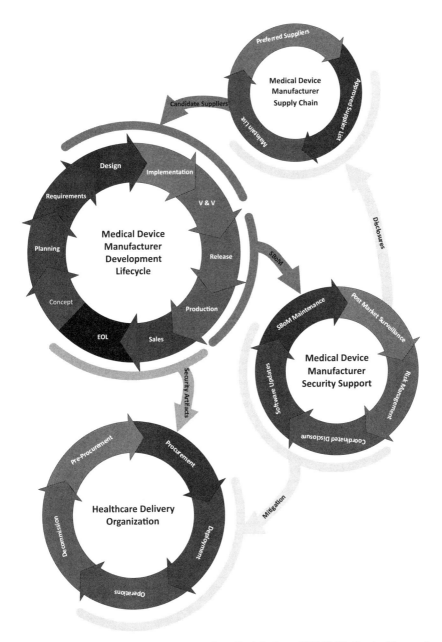

Figure 4.1 Secure development lifecycle of medical devices. (©H-ISAC. Used with permission.)

4.2 Secure Lifecycle Diagram Overview

In Figure 4.1, the four rings represent the entire secure lifecycle of a medical device as it applies to the main stakeholders—MDMs and HDOs—as well as the relationship between them. Each separate ring progresses asynchronously through its activities. These activities are oriented towards specific purposes, and in some cases, they inform or are themselves informed by activities in the other rings, as indicated by the arcs and arrows interconnecting the rings. Note that one of these rings is performed by the HDO (discussed in Chapter 7), while the other three rings—our focus in Chapters 3 through 6—are performed by the MDM.

The largest ring in the diagram, MDM Development Lifecycle, is probably the most familiar: it represents establishing the viability of the project with business case, market forecasts, schedules, and budgets; if successful, setting off on the development path; if successful, releasing the device into the market; then starting the whole process over with a new product. But many things have changed.

The first and easiest change we must now adopt is the concept of the ASL, governed by the topmost circle, MDM supply chain, as discussed in Chapter 3. ASLs are not a new concept, but historically they have been hardware-oriented and quality-focused, used for controlling sources of electronic parts and other components that met our needed functionality and tolerances and were based on our organizations' past experiences of interacting with a particular supplier. Now the process also includes TPSCs—libraries, frameworks, and operating systems—and suppliers are graded based upon those same traditional criteria plus a new addition: the supplier's approach and alacrity in addressing security issues.

The MDM security support ring addresses ongoing activities surrounding the postmarket surveillance of the device, as well as its hardware components and TPSCs. Disclosures from postmarket surveillance activities provide the data which helps drive the new ASL process (Chapter 3) as well as the secure update process for devices in the field (Chapter 6). In addition to creating timely awareness of security issues, security support processes also drive the mitigation and patching strategy employed in partnership with HDOs.

Ongoing cybersecurity surveillance is a new concept for traditional MDMs. Having to support the security posture of a product for its entire supported life is foreign and perhaps even slightly scary for some. Besides acquiring the expertise needed for these new activities, there is the question of who is going to pay for the new costs? Postmarket surveillance is rapidly becoming the new normal, an expectation of both customers and regulators. Yet few discussions about the costs and maintenance burden that will have to be passed on to HDOs are taking place.

Additional design considerations arise from the expectation that MDMs will support HDOs for the entire life of the medical device (as part of the HDO lifecycle), including supporting the HDO's decommissioning scenario (which is different from and independent of the MDM's EOL scenario). This means including functionality that securely implements a factory reset or data purge features in the device, which clears patient data (both PHI and PII), acquired values, settings, access control lists, user credentials, and shared secrets such as network or Wi-Fi credentials.

Possibly the most important part of this new interrelationship between MDMs and their customers is the security artifacts arrow leading from the MDM development ring to the HDO ring.

HDOs now routinely expect MDMs to deliver security artifacts as pre-sales deliverables in addition to those enforced through evolving regulatory guidances provided by agencies such as the FDA. Commonly requested presales artifacts are the MDS[2] and SBOM, but may also include other security contract obligations (all discussed in detail in Chapter 5).

Throughout this book, we refer to this four-rings diagram both to help the reader remain oriented and to ground each cybersecurity activity discussed within the overall lifecycle of secure medical devices. The present chapter focuses on the cybersecurity aspects of what is probably the most familiar ring, the MDM development lifecycle, from concept through release (AKA transfer to manufacturing).

4.3   Threats vs. Vulnerabilities

But before we dive in, some clarification on terminology and concepts (Figure 4.2). The cybersecurity industry frequently uses the industry-specific

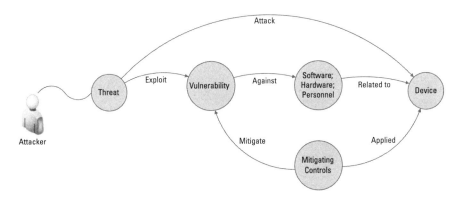

Figure 4.2   Applied terminology.

terms vulnerability and threat as if they were interchangeable (which they are not), which leads to confusion when identifying tools and techniques. This, in turn, leads to fuzzy-headed thinking about how to approach security, especially during the system design phase. This is due to the prototypical cybersecurity environment (e.g., the general IT environment) being a live, in-the-field IT installation that is exposed to a threat because of an inherent vulnerability. In this scenario, the practical distinction between the two terms amounts to a finely split hair, so the frequent conflation, although incorrect, is understandable. However, when applied to the design phase of a medical device, they are not even close to the same thing.

If the cybersecurity design of your device is based on the specific threat vectors and attack techniques known to be in use during the development timeframe, your device will be behind the security curve—insecure by the time it reaches the field.

Take ransomware as an example. The first ransomware appeared in 1989 and was distributed via floppy disks. Compare those humble beginnings to the ransomware of today, with so many variations on the theme, and so many vulnerabilities being exploited to infect the various platforms. What mitigations could have been implemented in 1989 that would still have any validity today? And yet a thirty-year device lifespan is not uncommon in medical devices.

During the design phase of a medical device, a requirement such as "The medical device shall not be susceptible to ransomware," (i.e., protecting against a specific threat) initially sounds like a pretty good idea! Who wants their medical device to fall victim to ransomware? But is that requirement actionable? No, it is not. There are many different types of ransomware [2], however, most literature focuses on what the ransomware does (encryption of files, locking the user out of the system, wiping the hard disk, etc.). Few studies focus on how the ransomware arrived in the device. Common infection vectors include [3]:

• Links in email or social messaging;

• Drive-by infections from a compromised website;

• Pay-per-install infections from botnet-compromised systems;

• Intrusion by exploiting vulnerabilities in the operating system or services, such as malformed server message block (SMB) packets and remote desktop protocol;

• Wormable intrusions that exploit network vulnerabilities;

• Unsecured local communication ports, such as USB [4].

So how does knowing any of this inform the medical device design? Short of removing all forms of communication from the medical device, there is little in the way of informing any design actions (or inactions) that need to be taken as a result of such threat-based requirements. Threat-based requirements do not answer any design questions, such as:

- Should hardcoded credentials be stored in an off-die external flash memory?
- Should a secure element be included in my design?
- Does the remote desktop securities (RDS) connection need additional layers of protection? What about for data traveling via Wi-Fi?
- Precisely what activities are needed to inform and guide design work?

The common theme is that a threat relies on the existence of vulnerability that can be exploited. However, preventing burglary is not the kind of job that cybersecurity requirements do. Our job is to secure the device against the burglar's most viable attack vectors: by buying better doors, installing better locks, and using them properly to lock all the doors—a task that generally starts with understanding where all the doors are.

A vulnerability is a weakness in design or implementation and a threat is the potential to exploit that vulnerability to achieve a goal. A weakness in the design cannot be exploited because it doesn't exist yet—it is just a concept developed during the design phase. Consequently, there is no threat to the design of a device (until that device has been built and is shipped).

Here are some further examples of key terms used in this chapter:

- Design vulnerabilities:
  - No authentication for a user to access the device;
  - No integrity checking of a data stream;
  - No encryption of PHI.
- Implementation vulnerabilities:
  - Didn't enforce the length of an input buffer;
  - Utilized intrinsic functions;
  - scanf() and fscanf() format strings didn't provide a field width for the %s string.
- TPSC vulnerabilities:
  - Heartbleed;
  - Urgent/11;
  - Sweyntooth.
- Threats:
  - Ransomware;
  - Worms;
  - Denial-of-service (DOS);

- Physical and side-channel attacks.
- Exploits:
  - Cryptolocker (ransomware);
  - NotPetya (ransomware);
  - Zcryptor (ransomware);
  - SYN flood (DOS);
  - Sasser (worm).

These concepts interact as in the following example: at design time, a design vulnerability is introduced, such as allowing access to the lowest levels of a medical device via a communications medium. This occurred because access to this communications medium was left open, with no authentication. At the time of design, all we can say is that we have a design vulnerability (that should be fixed). Once that vulnerability is in the field, it could be exploited in numerous ways, including by threats (ransomware, worms, botnets, etc.) and used for various purposes (e.g., extraction of intellectual property, extraction of PHI, compromise of design performance, just to name a few of the risks).

This is a different approach and way of thinking about the problem, and for people with IT backgrounds accustomed to dealing with deployed systems, this feels incorrect (and, for their environment, it is). But it is the correct approach for design developers. This is a very good thing because during the design and planning phases, trying to address threats is not actionable.

Fortunately, during the design phase, vulnerabilities can be identified in a formal, repeatable process that has been codified in various tools. (For further discussion of vulnerability assessment tools, please see the design phase section).

Vulnerabilities can be created or introduced not only during design, but also during implementation (more on that later) and through TPSCs. When distinguishing between vulnerabilities and threats, a TPSC is a special case because it is both a product already in the field and a component or subcomponent of the device. Therefore, it may have both vulnerabilities and threats—think for example of a commercial operating system you are using. This is one reason we treat TPSCs differently during development, more in line with how IT security is performed: we look for disclosed vulnerabilities as well as known threats to these components during a vulnerability assessment. TPSCs are discussed in greater detail in subsequent sections and chapters.

4.4   Development Lifecycle: Concept Phase

During this early part of the medical device development, the development team needs to consider all of the roles that interact with the medical device and their associated use cases.

Taken as a set, these use cases are constraints upon the security mitigations that can be considered, plus a set of potential vulnerabilities that may need to be mitigated.

Examples of these are:

- The user needs the functionality to copy PHI/PII from the medical device to a USB flash drive;
- The medical device needs to allow the user to attach their own keyboard via a USB connection;
- The medical device will utilize a Bluetooth low energy connection to any smartphone to communicate information to a cloud service;
- The medical device does not have an advanced user interface to authenticate a user.

Considerations should include such items as:

- User authentication;
- Decommissioning controls;
- Connections to other trusted entities;
- Examples of situations when break glass emergency access might be necessary for critical operations, where user-to-machine authentication can be bypassed.

Early proof of concept (POC) work can have an undue influence on the technical nature of the final device. To illustrate this, consider a small startup MDM that has limited funding and therefore selects low-cost components during proof-of-concept that cannot be secured. They expect to be able to leverage device-specific software in the final production version of this device. Porting their code to a new platform is expensive and time-consuming, and wasn't accounted for in their cost-projection model. Now the startup faces a sunk-cost scenario that it wasn't anticipating.

Early POC systems should utilize components that are capable of being secured, even if they aren't required for the POC version of the device. Components should include such features as:

- Cryptographic accelerators for AES, SHA, HMAC, CMAC;
- Secure boot;
- Secure key storage;
- Antidecapping measures;
- Physically unique functions (PUF);
- Virtualized security CPU;

• Side-channel power monitoring mitigations.

While these security features may not necessarily be active in the POC build, working with the same components in early builds will pay dividends when transitioning into the development of the real medical device. It can also smooth the way through security aspects of investigative device exemptions and investigative review boards.

4.4.1   Incremental Improvements and Secure Development

A common approach for existing MDMs is to make minor changes to an existing medical device to add a new feature or enhancement. This method of minor upgrades to a device has been an inexpensive way to attempt to keep an older device relevant and marketable while reducing engineering costs and traditionally keeping the regulatory hurdle low. However, great care should be taken before deciding to continue with this approach. Not only are 510(k) and substantially equivalent device submissions coming under increasing scrutiny for cybersecurity features specifically, but also existing medical devices may not have been designed with electronic components, operating systems, or application software that is capable of being secured against today's known threats.

The topic of legacy designs is a very difficult one because there are no easy fixes. The best course is to allow these insecurable designs to reach their end of life and be replaced with a newer platform capable of being secured. This will not often be a popular topic to discuss with senior management for the next few years of transition. They have likely participated in this model of continual improvement for a very long time, and they may be reluctant to let it go. However, not only is this a more responsible approach necessary for reducing exposures to patient and business risk, we anticipate that regulators and standard bodies will increasingly drive it to become normative.

4.5   Development Lifecycle: Planning Phase

Often, cybersecurity professionals from other industries are astonished at the lack of security-related planning incorporated into most medical device development projects. All too often, security practices in embedded device development lifecycles look like this:

1. Someone in the development team makes a list of all the security concerns they can think of.

    The persons carrying out this activity may or may not be security professionals with working awareness of the contemporary threat landscape; and they may or may not use rigorous, well-vetted, industry-accepted techniques and rubrics to identify and classify these concerns.

2. These concerns get included in the project as product security requirements.

3. Security is ignored until the first release candidate is ready.

   At this point, a third-party penetration tester is hired to attack the device and provide the MDM a report of all the security mistakes that were made.

4. This is way too late in the product development cycle to do anything about them, at least for reasonable effort and costs, and market pressure is already building. Many (or most, or all) of the device's vulnerabilities continue to be ignored (because no manager wants to admit they messed up), and the product is released.

Security should be treated like any other activity carried out during development. Every other part of the development lifecycle is planned for, scheduled, tracked, and documented. Security should be no different. This wholistic practice allows for coordination between teams, accurate tracking of project progress, realistic assessment of costs and resource needs, and smooth integration of subsystems for an overall high-quality product.

With cybersecurity aspects of medical devices becoming more and more regulated, it is no longer permissible for MDMs to take such a lax approach to security planning. A well-formulated security plan that breaks down the phases of the development project, lays out the security activities to be performed in each phase, and provides for their traceability and verification, is indispensable to securing approval in key markets worldwide, including in the United States, the European Union, France, Canada, Japan, Australia, and China.

4.5.1   Security Goals

The security plan can also be a good location to express the security goals of the project. These are not requirements, but rather overall high-level goals for virtually any type of development project. High-level goals for other considerations in development (such as the voice of customer, voice of business, etc.) are usually expressed in a separate document, and this can be done with security as well (i.e., voice of security). However, security's high-level goals are usually small in number, and therefore a voice of security document would be quite short. Locating them as introductory or background material in the security plan may be a more workable alternative than creating a separate document. More about the planning and content of the security plan can be found in Chapter 8.

4.6   Development Lifecycle: Requirements Phase

At first glance, it may seem that cybersecurity requirements for medical devices are obvious. Certainly, the CIA triad of "Confidentiality, Integrity, and Availability" seem like good starting points and they are; but the need for more granular levels of detail quickly becomes evident.

Depending upon the level of abstractions implemented in your document tree, the highest levels (e.g., system requirements) might just stop with the inclusion of the CIA triad, which could then be traced back to as the fountainhead for lower-level requirements, such as those defined in device-specific requirements documents and interface control documents.

This section of the book is designed to help the reader with the task of sorting out the wheat from the chaff in defining the content of cybersecurity requirements. (Guidance for structuring requirements documentation is found in Chapter 8.) The first step in achieving that is to remove all references to threats. As previously discussed, threat-based requirements are those designed to counter specific threats or even specific categories of threats, when instead, the focus should be placed on vulnerabilities.

Vulnerabilities can be created or introduced in three ways during development: by design, during implementation, or through TPSCs. The traditional approach to securing a medical device is to break these areas down into the well-known security triad: confidentiality, integrity, and availability (CIA). Examples of these include:

- Confidentiality:
  - Of the patient's sensor readings;
  - Of the medical device settings;
  - Of the patient's PII;
  - Of the firmware upgrade image used to upgrade the device while in the field;
  - Of the MDM's intellectual property;
  - Of the command-and-control functionality exposed via communications mediums;
  - Of user or network credentials.
- Integrity:
  - Of the essential performance of the medical device;
  - Of the patient's sensor readings;
  - Of the patient's medical device settings;
  - Of the patient's PII;
  - Of the firmware upgrade image used to upgrade the device while in the field;

- Of the command-and-control functionality exposed via communications mediums;
- Of device-specific data like calibration settings or safety limits.
- Availability:
  - Of the essential performance of the medical device;
  - Of the patient's sensor readings in the expected time frame;
  - Of the patient's medical device settings;
  - Of the command-and-control functionality exposed via communications mediums.

In the health care industry, great emphasis has traditionally been placed on the confidentiality of data (e.g., through national laws like the HIPAA Breach Notification Law in the United States or GDPR in Europe). However, there is also an implicit ranking of the CIA triad from the MDM's perspective. MDM security activities typically prioritize ICA, where integrity is considered paramount while the availability of the device is not seen as quite so important. This can be seen in the handling of many types of alarms in existing medical devices and proposed designs, where normal treatment is suspended or reduced to a safe fallback mode of operation while the user is informed of the alarm. HDOs, meanwhile, may be more interested in being assured of the availability of the device's essential clinical functions and the integrity of the both the device and any connected devices within the HDO's network, with confidentiality seen as slightly lower priority in the face of immediate care delivery needs—AIC. Given these slightly different emphases in priorities, determining cybersecurity requirements requires careful evaluation based on the specific nature of the medical device.

At its simplest level of categorization, there are two main device architecture types: the embedded PC and the custom hardware with microcontroller.

For the embedded PC, detailed design objectives may include some or all of the following:

- The commercial off-the-shelf operating system shall be hardened to remove all unused functionality, including:
  - Disabling the USB ports or, if not possible, disable all but the anticipated functionality supported (e.g., disable USB autorun function);
  - Disabling all unused network ports and protocols;
  - Disabling all unused functionality and processes;
  - Inhibiting all hotkeys;
  - Automatic encryption of the mass storage device (hard drive or SSD);
  - Disabling automatic updating of the operating system.

Note that while hardening baselines such as those provided by the OS manufacturer or the Department of Defense (DOD) are good standards for controlling this effort initially, some minor modifications of a given baseline may be necessary to support specific device/system use cases, like:

- The operating system shall be placed into a kiosk mode or equivalent such that only the medical device application can be executed.

- The BIOS setup shall be protected from alteration.

- Specific environmental functionality shall be supported. Considerations should include functions such as:

  - Inclusion in an environment's active directory.

  - Automatic identification to the intended device installation site's asset management system.

  - Installation of third-party host intrusion detection system (IDS) or intrusion prevention system (IPS). A third-party Host IDS/IPS offers the advantage of avoiding device-hosted signature databases that need to be updated and other disadvantages of traditional antimalware technology (see Chapter 10 for more).

  - Installation of commercial antimalware (if appropriate for the device type). Note that including commercial antimalware tools introduces a new set of considerations, including the ability to update the virus signature database, the possibility of false-positive detection of valid medical device software on the PC, and the possibility of exposing the device directly to the internet or local intranet to retrieve signature updates. Depending on the configuration, other challenges may also require consideration, such as handling how and where a detected virus notification will occur, especially if the OS is in kiosk mode during normal operation.

- The application shall support secured functionality to grant authorized users' access to underlying operating system functionality not typically granted to a normal user, such as joining a domain or connecting to a Wi-Fi access point.

- End-of-life functionality shall be supported. An example of such functionality is the deletion of all data at rest, as is typically provided by a factory reset.

- Updating the operating system and updating of the application shall be supported.

For custom hardware with microcontroller or commercially available single-board computer (SBC) designs, where we are not trying to restrict a flexible commercial operating system into a single-purpose machine, detailed requirements may include some or all of the following:

- Secure coding conventions shall be defined and followed for each of the computer languages utilized in the system.

- Secure coding conventions shall be enforced via static analysis of the generated code.

- Secure boot functionality shall be implemented in your microcontroller to ensure the execution of unadulterated code.

- Protection of confidentiality, integrity, and availability shall be applied as widely as possible throughout the design and implementation of software and hardware for the device and system.

- The implemented security shall not be brittle, meaning that it should allow and/or detect outlying conditions and events present in nonnormative conditions without negative impacts on the security.

- Security mitigations shall be interlocking with other security mitigations where violations to one have detectable consequences in the other (AKA defense in depth).

- Nonvolatile memory shall be partitioned to ensure that the security mitigations and intellectual property can be contained on the same die as the CPU (as opposed to external nonvolatile memory).

- CPU-provided memory protection shall be utilized in the microcontroller to partition processes and any potential data storage into separate areas, as well as to designate which memory addresses can contain executable code.

- Exception handling shall be configured not only to detect failures but also to log them, notify the user, and continue to deliver the essential performance of the medical device.

- Processes shall never be blocked due to resource exhaustion (e.g., from dynamically allocated memory). Resources shall be statically allocated to processes.

- General-purpose operating systems shall be avoided in favor of smaller schedulers or bare-metal round-robin execution (all extra functionality in operating systems increases the possible attack surface).

- Due to the limited scope of default, general-purpose mitigations, the use of mitigations implemented in communications mediums shall not be considered adequate, except as part of a defense-in-depth strategy along with other mitigations.

- Modes of operation shall be considered in the system, especially where nonclinical functionality is performed. The usual cases for this are production-line configuration and testing support in a device/system. While manufacturing functionality may need to be active in a device/system that is currently in the process of being manufactured, this functionality should be removed or at least disabled in a device/system once it leaves the manufacturing environment.

- The system design shall ensure that in case of a compromise of a single device, the information learned cannot be used to facilitate additional attacks against other similar systems, especially larger scope attacks. Typically, such exposures are the result of default credentials, a shared secret present in all devices, or hardcoded credentials.

- Hardware root-of-trust functionality shall be preferentially implemented over software-only solutions. This includes key and shared secrets storage as well as encryption/decryption acceleration.

- There shall be encryption of all critical data. This includes data in motion, data at rest, and firmware updates.

- Encryption shall, at a minimum, utilize symmetric encryption of not less than 128 bits in width with a preference of 256 bits in width.

- Data integrity shall be ensured via digital signatures, cryptographically strong hashes, or message authentication codes, such as HMACs and CMACs.

- Data authorization shall be ensured via access control lists or similar structures.

- There shall be authorization control of devices in the system's communication network.

- Integrity-secured logging shall be used to ensure the integrity of the content in the log, as well as the integrity of the order of the records added to the log.

- Keys and shared secrets shall be stored in a secure manner (i.e., not in plain text nor in easily accessible memory).

- Keys and shared secrets shall be device-unique and ephemeral.

- Apps targeting commercial operating systems (e.g., Windows, iOS, Android) shall be obfuscated.

- Currency of received data shall be monitored for significant delays and if present, reported to the user.

- Handling of shared secrets such as encryption and authentication keys shall be performed in a nonpredictable bit/byte manner to prevent the possibility of power monitoring side-channel attacks that would expose these values.

- Comparisons of values utilized for security mitigations shall be performed in a time-agnostic manner, such that comparison success or failure occupies identical periods of time.

- Intrinsics such as memcmp, memcpy, or strcmp shall not be utilized for handling of any shared secret.

- Compiler optimizations shall be disabled around the code that is used as an alternative to intrinsics such as memcmp, memcpy, or strcmp.

- In-the-field firmware updates shall be supported.

- Firmware update images shall be protected from exposure of image contents (i.e., by encryption).

- Firmware update images shall be protected from undetectable adulteration of the image contents and authentication of origin (i.e., by integrity checks via keyed cryptographic hashes).

- Firmware update images shall be protected from rollback attacks (i.e., by allowing advances in version numbers only).

- Firmware updates shall be authenticated (i.e., cryptographically signed).

- Interruptions to essential performance shall, at a minimum, be reported to the user.

Additional requirements and detailed designs will be created during the design phase based on the results of the Vulnerability Assessment (discussed below).

4.6.1 Safe Harbor vs Full Encryption

The HIPAA Safe Harbor provision is part of the HIPAA Privacy Rule. It applies generally to HDOs and to any MDMs whose products or services include handling of PHI. However, the HDO may contractually require HIPAA-compatible (or stronger) data protection to enable its compliance with the law and lower its risk exposure.

Safe Harbor is a method of deidentifying data by removing specific types of information in PHI to prevent reidentification of the PHI. This means that the following types of data need to be removed from the PHI dataset to be considered deidentified by HIPAA (note: this is a legal position, and is not the same as truly being deidentified):

- Names;
- Social Security Number;
- Telephone, cellphone, and fax numbers;
- Email address;
- IP address;
- Driver license number;
- Any uniquely identifying numbers or addresses, including network/Wi-Fi/Bluetooth/MAC addresses and IMEI addresses from cell phones;
- Geographic locators (physical addresses, longitude/latitude coordinates, etc.);
- Only the first 3 digits of a zipcode are permitted;
- All dates except for the year, and all birthdates (including the year);
- Medical record number;
- The member ID on a patient's health insurance card;
- Device serial number;
- Birth certificate number;
- Any account numbers, including bank;
- Anything to identify a vehicle including VIN numbers and license plate;
- URLs;
- Clear, easily identifiable photograph;
- Biometric data, including fingerprints, voiceprints, retinal images, gait, and unique cardiac identifiers.

Every MDM should be aware that there have been multiple examples [5] (far too many to list) of anonymized data being re-identified, going even so far as to track secret service agents [6]. Where there used to just be HIPAA, now there are many laws for protecting information and the privacy of the patient, and many of these laws have serious financial consequences for violating them. Lastly, HDOs are now increasingly including language in contracts that specify that all data needs to be encrypted, whether at rest (database, buffer, etc.) or in motion (i.e., transmitted via wired or wireless communications). It is therefore recommended that encryption for all data be strongly considered for your medical devices, along with strong cryptographic primitives (i.e., hashes) to ensure the integrity of the data as well.

4.7 Development Lifecycle: Design Phase

Identifying, assessing, and managing vulnerabilities during the design phase of a project can be difficult. Little in the way of clear guidance has been made available, as few academic studies have been performed on identifying and preventing design-phase vulnerabilities. Even the concept of what engineers should seek to identify and mitigate during the design phase (vulnerabilities, not threats) is not widely understood.

This is one of the largest differences between MDMs and normal MIS/IT security. IT security teams use tools and techniques to detect intrusion attempts and breaches (threats), whereas the MDM uses tools and techniques to detect vulnerabilities and prioritize their mitigation. As discussed in detail before, this is a subtle and often misunderstood difference, with significant impacts on how product security is practiced by an MDM.

4.7.1 Design Phase Activities

Utilizing threat tools (an established but confusing name) at design time to expose areas of a proposed design that need mitigating controls is exactly the process detailed in this section. The challenge inherent in this approach is that development engineers must apply tools and techniques developed and used by MIS/IT security defenders in the field. These are different domains that need slightly different treatments when applied during the design of the medical device. At the most basic level, it is the MDM's goal to remove potential cybersecurity exposures that, if exploited, would result in negative impacts on a patient (risk of harm) or business (risks to business model). However, after a product has been released into the field, threat modeling can inform experts in emergency preparedness and planning. At that point, threat modeling may be able to provide guidance as to where changes in the device utilization, configuration, or environment could be made to reduce the possibility of an attack succeeding.

What is needed during a medical device's design phase is a formal approach that minimizes the participation of a cybersecurity SME in the process, as well as a nonambiguous method to logically define the system being designed that will constrain and define the list of potential vulnerabilities as derived by decomposition of the defined system. This list of potential vulnerabilities then needs to be screened for false positives and the remainder prioritized in consideration of the use cases for the particular medical device system. The prioritization process also needs to differentiate risks to users (device safety and efficacy) from business risks (threats to the business model) that the proposed design could pose to its MDM if taken as-is into full production. Finally, this design phase vulnerability detection tool needs to be as automated as possible, to allow for speed of design, redesign, and reuse in other development projects.

The closest currently available tool for achieving these goals is the STRIDE threat modeling tool from Microsoft. As discussed before, threat may not be correct here, but the name is well established and we can't turn back history!

STRIDE stands for:

- Spoofing identity;
- Tampering;
- Repudiation;
- Information disclosure;
- Denial of service;
- Elevation of privilege.

In the original, IT-focused definition of STRIDE, its six areas of concern were specifically defined to address IT risks. For example, spoofing identity was directed towards spoofing a user's identity. When focused on medical devices, we have to be assured not only of the user's identity (including a variety of user roles like patient, clinical, technical, or remote users), but also of the identity of any other connecting device (i.e., machine to machine).

Enlarging the definition of these terms works well with the existing STRIDE model, so here are the redefinitions of those same six areas for use with an embedded medical device system:

- Spoofing identity:
  - Utilizing another user's authentication;
  - Utilizing a different device or IT system to communicate into the system.
- Tampering:
  - Malicious or unauthorized modification of data at rest;
  - Malicious or unauthorized modification of data in motion;
  - Malicious or unauthorized modification of program executable.
- Repudiation:
  - The ability to deny performing an action.
- Information disclosure:
  - Exposure of information to an unauthorized user;
  - Exposure of data at rest;
  - Exposure of data in motion;
  - Exposure of program executable.
- Denial of service:
  - Negative impacts on medical device system availability and reliability;
  - Preventing valid users from authenticating;
  - Blocking or delaying communication mediums carrying data in motion.
- Elevation of privilege:

- Unprivileged user gains privileged user's access (a subcase of spoofing);
- Unauthorized connection into the medical device system.

These are then applied to the set of primitives, as shown in Table 4.1. The example mitigations for each of the STRIDE classifications are:

- Spoofing identity:
  - Perform authentication;
  - Compromise cryptographic hashes;
  - Protect secrets utilized in authentication (i.e., point-to-point connections);
  - Utilize highly ephemeral secrets in authentication;
  - Utilize third-party authentication, such as PKI.
- Tampering:
  - Perform authorization;
  - Compromise cryptographic hashes;
  - Message authentication codes;
  - Digital signatures;
  - Secure boot.
- Repudiation:
  - Digital signatures:
  - Audit trails;
  - Cryptographic hash chains;
  - Forensically significant logging;
  - Logging of key events, including security events.
- Information disclosure:
  - Perform authentication;
  - Perform authorization;

Table 4.1
Applying Threat Modeling

	Process	Data Stores	External Entities	Data Flows
S	✓		✓	
T	✓	✓		✓
R	✓	✓	✓	
I	✓	✓		✓
D	✓	✓		✓
E	✓		✓	

- Compromise encryption;
- Disclose secrets, such as encryption keys;
  - Utilize highly ephemeral secrets, such as encryption keys.
- Denial of service:
  - Perform authentication;
  - Perform authorization;
  - Filtering of communications;
  - Throttling of communications;
  - Monitor quality of service of communications (i.e., stale data);
  - Ensure proper bandwidth for the essential performance of the medical device.
- Elevation of privilege:
  - Provide multiple methods of authentication/authorization;
  - No single layer of authentication (i.e., avoid the binary concept of authentication).

In the intended utilization of STRIDE, an engineer would lay out the data flow diagram (DFD) of newly developed software using the Microsoft threat modeling tool (TMT). An example is provided in Figure 4.3.

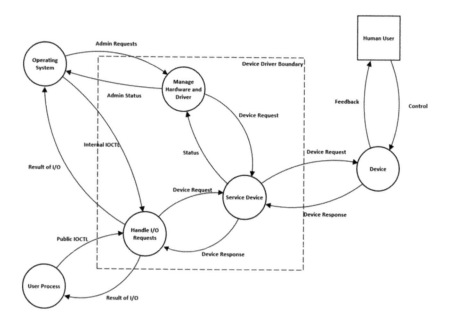

Figure 4.3 Sample TMT output using software data flow

The TMT would then automatically apply the STRIDE rules to this DFD and decompose the diagram into a list of potential vulnerabilities, output into a spreadsheet.

At this point, a few items to note. First, STRIDE was initially designed to be a software-only tool, and thus it would be assumed that the software would have already been written before STRIDE was performed. In other words, it was not initially well-suited for design phase work. Second, the list of potential vulnerabilities is as complete as the DFD entered. There is no concern about the level of completeness being represented in the list of potential vulnerabilities. The obvious level of completeness being displayed in the DFD will be the same as represented in the list. Third, the DFD is formally decomposed into the list of potential vulnerabilities automatically, without regard for the security skill set of the engineer.

But this software-focused tool is very flexible, and with some small alterations can also be used during the design phase of development to create a list of potential vulnerabilities in the system-level design being proposed. Figure 4.4 provides a second example from the Microsoft TMT, this time representing a system-level topology with logical data flows.

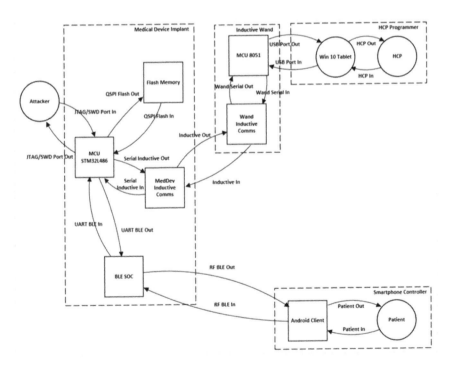

Figure 4.4   Sample TMT output using system-level topology with logical data flows.

At this system-level representation, a list of potential vulnerabilities is automatically created from the data flow diagram. The TMT accepts as input the overall system topology and any particular device details, including schematics of major components, data stores on smartphones, and the nature of utilized communication mediums. However, due to the abstract nature of STRIDE (e.g., it doesn't know the difference between a JTAG connection as compared to an ethernet connection), the list of potential vulnerabilities usually has about 30% false positives and duplicates that will need to be removed through manual inspection.

4.7.2 Introduction to Vulnerability Scoring

Now we have this list of potential vulnerabilities in our design. This list can easily grow into the hundreds, including false positives as well as duplication of possible vulnerabilities. We could attempt to mitigate every listed potential vulnerability. However, that approach would most likely waste time and money as not all of these potential vulnerabilities are applicable due to the use case of the medical device. Also, not all potential vulnerabilities can be exploited.

As an example, a design might use a BLE radio to broadcast the battery status of your medical device. Usually, this is information that nobody cares about exposing to anyone. However, if that same BLE interface could modify the amount of therapy delivered to a patient, that is a completely different use case that requires mitigations in place to protect the essential performance of the medical device.

If your project has an unlimited budget and schedule, then by all means, please investigate and mitigate all of the entries in the list of potential vulnerabilities (we're sure that such a company exists somewhere, but we have yet to find it)!

The reality is that MDMs have limited time and budget to identify and mitigate the vulnerabilities that matter to the product. Specifically, we need a metric, created by a rubric, that allows a quick and easy examination of each potential vulnerability's impact on confidentiality, integrity, availability (the CIA triad) including exploitability and severity. To achieve this level of prioritization, a second operation following the potential vulnerability list generation must be performed—one that:

- Scores the list of potential vulnerabilities;
- Eliminates the false positives and duplicates (inherent in the abstract nature of STRIDE).
- Delineates the exploitability.
- Estimates the severity.

- Flags each potential vulnerability for its potential to impact the safety and efficacy of the device for end-users. Note, however, that any vulnerability that is not flagged for impacts on safety and efficacy should still be considered as a vulnerability that threatens the MDM's business model.

These impacts to the CIA triad need to be tempered by the severity and exploitability that are possible. In the case of vulnerabilities that pose the possibility of patient risk, the security engineer does not have the tools, experience, or education to make the judgment call if this is an impact on safety and efficacy. Instead, flagging a vulnerability indicates that this item needs to be introduced into the company's existing risk management system (RMS) for review.

A good example of why cybersecurity risks that potentially impact safety and efficacy are referred into existing risk management processes is illustrated by the temporary cessation of the medical device's essential performance. Let's say that the engineer or security professional identifies a vulnerability that could allow an attacker could disrupt the functionality of this device for five minutes, and so flags it as a patient risk. This is then transferred to the RMS, where it is reviewed and determined that, either

1. This isn't a risk to a patient (e.g., the device is a personal insulin infusion pump);
2. It is a risk to the patient (e.g., the device is a ventilator).

Staff engineers and security professionals should not be performing tasks that the risk management process exists to control.

So how do we know which of these potential vulnerabilities matter to the security of our device? Up to this point, we have been using the TMT tool and thinking like an engineer. Now we have to start trying to think like a hacker. As we move to score the list of potential vulnerabilities, being aware of common attacks is necessary for scoring the ease of attack (AKA exploitability) of the vulnerability. A good example of this, and one that we hear frequently is, "Since the software source is compiled, no one can read it!" This is not true; however, it remains a surprisingly widespread belief. If you are unaware of such malicious capabilities, you cannot accurately score the exploitability of a vulnerability.

The currently available rubric that can provide the necessary insights into design vulnerabilities is a modification of the existing CVSS v2 threat scoring rubric (discussed in detail below).

Following the application of CVSS to each of the potential vulnerabilities, we determine that any potential vulnerability with a score greater than 4 should be investigated with the intent of implementing a mitigating control. It is also at this point where any false positives are identified and removed from further vulnerability processing.

4.7.3   Mitigations

Mitigating controls are always partial solutions, never a complete solution to the vulnerability, and in some cases no mitigation is even possible. One such example is a denial of service (DoS) attack against a wireless connection, where so much radio frequency (RF) noise is blasted out with a high enough amplitude that the RF connection will not be available until the disruptive broadcast ceases. The only real mitigating control for this is probably already planned for in the design, namely, notify the operator of the device's inability to make or sustain a connection to the other device. This is not a perfect mitigation but given the reality of the communications medium, this may be the best the MDM can achieve.

Wherever possible, mitigations are created for each of the high-scoring vulnerabilities. Then the scoring process is repeated on each of these mitigated vulnerabilities, this time considering the impact of the implemented mitigating control. Outputs of this process inform the project's architecture control documents, requirements, design, risk management activities—in short, the remainder of the development lifecycle, as shown in Figure 4.5.

4.7.4   Vulnerability Scoring

Initial vulnerability scoring assumes that no mitigation exists for the vulnerability. Specifically, the list of scored vulnerabilities is what guides the priority and extent of efficacy for each of the required mitigations.

After the design and selection of mitigations for each scored vulnerability, the list of vulnerabilities is rescored. In this second round of scoring, mitigations

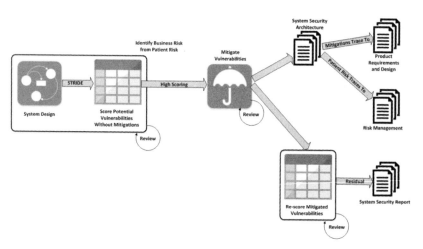

Figure 4.5   Cybersecurity activities during the development lifecycle

are included to produce a postmitigation score for each vulnerability. This second round highlights whether all mitigations are sufficient as designed, or whether additional mitigations are needed for any residual vulnerabilities.

If no further mitigations are possible then these high scoring vulnerabilities are considered residual vulnerabilities that will be present in the production version of the medical device. The vulnerability assessment (VA) process is included in the secure development cycle as shown in Figure 4.5.

Note that the VA process we're describing does not (and logically cannot) address vulnerabilities that may be introduced via implementation. It is solely focused on applying a rigorous formal method to expose areas in the design itself that necessitate mitigating security controls.

4.7.5 Scoring Rubrics

It is critical to apply the right scoring tool to the appropriate point in the product lifecycle: vulnerability assessments are applied during design; implementation vulnerability assessment during development; and threat modelling for a product in use in the field. There are many documented criticisms of existing vulnerability scoring rubrics, and while these criticisms may be valid, they are often made in the context of finished, installed devices in the field. For such devices, cybersecurity assessments should be made against security risks, which are scored and mitigated differently than vulnerabilities. Our focus for this section is on vulnerability assessment rubrics used to evaluate the security of a proposed design. Until a universally accepted standard for scoring design phase vulnerabilities is created, we recommend modifying common vulnerability scoring system (CVSS) v2.

The CVSS is defined by FIRST's CVSS Special Interest Group [7]. FIRST is the Forum of Incident Response and Security Teams; as indicated in their name, this group is comprised of responders to security incidents, primarily breaches of postmarket products and deployed infrastructure. Consequently, CVSS is segmented into three major groupings of metrics: base metrics, temporal metrics, and environmental metrics.

This organization system is indicative of FIRST's postmarket viewpoint. In particular, the Temporal group and the environmental group consist of metrics that have no relevance to scoring a vulnerability during the design phase of a medical product. By contrast, in the base group, all of the metrics are relevant to vulnerabilities discovered during the design phase of medical device development.

A metric that exists in the environmental group of CVSS version 2 (but not in version 3) is collateral damage. This metric is semantically equivalent to a more commonly used term in medical devices, severity. Severity represents the potential harm to users, as well as the potential risk to the device manufacturer's business model, which could result from a given vulnerability being exploited. It is the single most important metric when scoring a medical device design.

During the design phase, our purpose is to assess vulnerabilities and assign a score to create a meaningful overall score that includes severity (collateral damage from CVSS v2) as well as exploitability (base metric group), and measures their impacts to confidentiality, integrity, and availability (base metric group). Our final score must be one that prioritizes which vulnerabilities to mitigate in the device design. Therefore, a customized CVSS approach is suitable, and is already in use by several MDMs for premarket assessment activities.

4.7.5.1 Modifying CVSS v2

While CVSS v2 comes close to what we need for scoring design-phase vulnerabilities, there are several aspects to it that need to be extended and modified to support this effort. Several attributes are not valid and thus must be set to values that nullify their effect on the overall score. Table 4.2 shows the nullified attributes and their assigned values.

Note: In the following pseudocode of the rubric, "1" digits are left to represent the nullified attributes.

```
If Base Section completed
   Impact = 10.41 * (1 - (1 - ConfidentialityImpact) * (1 - Integrity
   Impact) * (1 - AvailabilityImpact))
   If Impact=0 then
      fImpact = 0
   else
      fImpact = 1.176
   ExploitabilityScore = 20 * AccessComplexity * Authentication *
   AccessVector
   BaseScore = Round_To_Tenths (((0.6 * Impact) + (0.4 * Exploitabili-
   tyScore) - 1.5) * fImpact)
If Environment Section completed Then
   AdjustedImpact = Round_To_Tenths (Minimum (10, 10.41 * (1 - (1 - Con-
   fidentialityImpact *
      ConfidentialityRequirement) *
      (1 - IntegrityImpact * IntegrityRequirement) *
```

Table 4.2
Nullified Values for CVSS v2

Attribute	Value
Exploitability	high (1)
Remediation level	unavailable (1)
Report confidence	confirmed (1)
Target distribution	high (1)

```
      (1 - AvailabilityImpact * AvailabilityRequirement))))
   AdjustedBaseScore = Round_To_Tenths (((0.6 * AdjustedImpact) + (0.4
      * ExploitabilityScore) - 1.5) * fImpact)
   AdjustedTemporalScore = Round_To_Tenths (AdjustedBaseScore * 1 * 1
      * 1)
   EnvironmentalScore = Round_To_Tenths ((AdjustedTemporalScore + (10
      - AdjustedTemporalScore) * CollateralDamagePotential))
If EnvironmentalScore = 0 Then
   OverallScore = BaseScore
Else
   OverallScore = EnvironmentalScore
```

4.7.5.2   Redefining CVSS v2 Attributes

Redefinition of the remaining (nonnull) CVSS attributes is necessary to better address the embedded device domain as compared to the original IT domain. Our recommended modifications are described below:

- Access vector:
  - *Local*: A vulnerability exploitable with only local access requires the attacker to be in physical contact with the device. Examples of locally exploitable vulnerabilities are USB ports, Serial ports, JTAG debugging ports, and so on.
  - *Adjacent network*: A vulnerability exploitable with adjacent network access requires the attacker to have access to either the broadcast or collision domain of the vulnerable software. Examples of local networks include local IP subnet, Bluetooth, IEEE 802.11, local Ethernet segment, BLE, and low-power radio communications, with a maximum range measured in hundreds of feet.
  - *Network*: A vulnerability exploitable with network access means the vulnerable software is accessible via a network stack and the attacker does not require local network access or local access. Such a vulnerability is often termed remotely exploitable. Wide area wireless networks (i.e., cellular radio) and broadcast radios would also be included in this category. In certain cases, a link to a smartphone could also be considered a network attack.
- Access complexity:
  - *High*: Extremely difficult to exploit, consisting of hundreds of man-hours and/or expenses running into the hundreds of thousands of dollars. Specialized access conditions exist. For example:
    - The attack depends on the victim performing several suspicious or atypical actions;
    - The vulnerable configuration is seen very rarely in practice;

- The vulnerability depends on a narrow window in time;
- Exploiting the vulnerability requires expensive tools.
- *Medium*: The access conditions are somewhat specialized. Examples include:
  - The attacking party is limited to a group of systems or users at some level of authorization, possibly untrusted;
  - Extensive information must be gathered before a successful attack can be launched;
  - The affected configuration is a nondefault configuration that is not commonly utilized;
  - The attack requires a small amount of social engineering that might occasionally fool cautious users;
  - The attack requires insider knowledge of the device's hardware or software design.
- *Low*: Specialized access conditions or extenuating circumstances do not exist. Examples include:
  - The affected product typically requires access to a wide range of systems and users, such as anonymous and untrusted connections with no authorization;
  - The affected configuration is default or ubiquitous;
  - The attack can be performed easily and requires little skill or additional information-gathering;
  - A race condition exists; however, it is a lazy one (i.e., it is an easily winnable race);
  - Distributed system gives attacker all the tools and/or knowledge to exploit the vulnerability;
  - Commonly available tools, such as those easily downloadable from the internet, can exploit the vulnerability.
- Authentication:
  - *None*: Authentication is not required to exploit the vulnerability;
  - *Single instance*: Authentication is easily defeated or uses a weak method for vetting;
  - *Multiple instances*: Authentication employs the industry's best practices for vetting the authenticity of the user or device.
- Confidentiality impact:
  - *None*: There is no impact on the confidentiality of the system.
  - *Partial*: There is considerable informational disclosure. Access to some data (including executable) is possible, but either the attacker does not have control over what is obtained, or the scope is constrained.

An example is a vulnerability that divulges only certain fields of data out of the set of what is available.

- *Complete*: There is total information disclosure, resulting in all data (including executable) being revealed. The attacker is able to read all of the data possible from this vulnerability (volatile or nonvolatile memory, data streams, external communications, data at rest, data in motion, etc.).

- Integrity impact:
  - *None*: There is no impact on the integrity of the system, including on the operation of the system.
  - *Partial*: Modification of some data (including executable) is possible, but either the attacker does not have control over what can be modified or the scope of what the attacker can affect is limited compared to what is available to this vulnerability. Some alteration of essential performance is possible. For example, part of the executable image may be overwritten but not all of the executable image.
  - *Complete*: There is a total compromise of system integrity. There is a complete loss of system protection, resulting in the entire system being compromised. Total control over the essential performance of the device is possible. The attacker can modify any data (including executable) on the target system.

- Availability impact:
  - *None*: There is no impact on the availability of the system.
  - *Partial*: There is reduced performance or interruptions in resource availability, reduced throughput or increased latency in responding to an event, or reduced performance where some essential performance is impact including missed communication periods. An example is a network-based flood attack that permits a limited number of successful connections to an Internet service.
  - *Complete*: There is a total shutdown of the targeted device, rendering the device's essential performance nonoperational. The attacker can render the resource completely unavailable. An example is a ransomware attack.

- Collateral damage:
  - *Low*: A successful exploit of this vulnerability may result in slight physical or property damage, including a slight loss of revenue or productivity;
  - *Low-medium*: A successful exploit of this vulnerability may result in moderate physical or property damage, including moderate loss of revenue or productivity;

- *Medium-high*: A successful exploit of this vulnerability may result in significant physical or property damage, including significant loss of revenue or productivity;
- *High*: A successful exploit of this vulnerability may result in catastrophic physical harm (such as death) or property destruction, including catastrophic loss of revenue or productivity.

4.7.5.2   Differences Between CVSS v2 and v3

Some further mention of CVSS v3 should be made, as it has changed in ways that don't work as well for design vulnerability scoring. FIRST released a revision to its CVSS rubric in June 2019, officially updating CVSS to version 3.1, primarily to better address threats in an MIS/IT system in a live production environment. Despite these updates, CVSS v2 remains widely used, especially for scoring vulnerabilities in medical device design. There are several reasons for this, but the most influential is FIRST's decision to remove the collateral damage attribute. Since this attribute is used to represent severity in design phase vulnerability scoring, CVSS v3 lost the single most important and consequential attribute of its final score. It is for this reason that design phase scoring has not adopted newer versions of CVSS.

A whitepaper containing a complete description of all the changes between CVSS v2 and v3 and their impacts on medical device design phase scoring can be found in the References section of this chapter [8].

4.7.6   Alternative Approaches to Scoring

Over the past decade, a large variety of scoring rubrics have been created. They are usually created to fit a specific niche need or industry, and all are written from the perspective of a released device—designed to assess threats rather than vulnerabilities. These rubrics usually fall into one of three approaches:

   A. Multiple subjective factors/attributes, all equally reflected in the final score;
   B. Multiple subjective factors/attributes, with each attribute being conditioned by a weighting factor before being calculated into a final score;
   C. Either (A) or (B) with the inclusion of a likelihood metric.

4.7.6.1   Likelihood

Because the medical device has not yet been created, there is no likelihood of it being exploited. This metric doesn't pertain to the context in which we are deploying the rubric. Attempts to assign a probabilistic rate of attack or exploit occurrence are futile, impossible to ascertain, except as a best guess regarding conditions that may exist in the future if the device were to be manufactured and sold according to the early-stage design in front of us. It relies on speculation and conjecture, in other words, and is not useful for identifying and incorporating appropriate security mitigations into your design.

Some rubrics try to conflate likelihood with other attributes, typically called out as metrics with names like:

- Attacker skill level;
- Attacker motive;
- Opportunity to exploit;
- Size of attack;
- Ease of discovery;
- Ease of exploitation;
- User awareness;
- Intrusion detection.

Most of these share the same basic issues with likelihood. Not only are they dependent on speculation and guesswork, but they are also focused on threats, not vulnerabilities.

Only one of these makes a reasonable contribution to medical device design scoring, and that is ease of exploitation. This is still quite subjective, but it does add some meaning to the final score. For example, if exploiting the vulnerability in question requires the active use of elaborate equipment in the same room as the device being attacked, it may be considered relatively more difficult to exploit because of the challenge of transporting and setting up the equipment in close enough proximity without being detected before the attack can be carried out. On the other hand, if the exploit can be carried out from anywhere with the click of a button, it may be considered relatively less difficult to carry out.

A good, representative example of how likelihood and related metrics are typically deployed by these rubrics is the National Institute of Standards and Technology's (NIST) risk determination rubric, where likelihood and severity are both equally influential upon the final score. Under this rubric, the assessor could determine that the vulnerability's severity is very high because a patient could die from this vulnerability being exploited. However, in a case where the previous version of the device has been on the market for many years without a known attack against that same design vulnerability, the assessor could rate the likelihood as very low. The resultant score would be low, and thus, the scoring rubric would allow the designers to ignore it.

Needless to say, this is extremely problematic, and could lead to scrutiny and delays in the (still evolving) market approval process! Either the device will not be approved for the market because the flawed rubric used for the design vulnerability assessment failed to flag the vulnerability and the design was submitted for regulatory approval without adequate mitigation, or worse, the flawed rubric will mislead reviewers, who then approve the device for the market, only to see it later become the means by which one or more patients is severely harmed due to this unlikely vulnerability being exploited when it matters.

Conversely, if you assume that likelihood is always high, then this metric simplistically magnifies severity and distorts the comparative outcome when identifying vulnerabilities in need of mitigation. Either way, the fundamental problem with these types of rubrics is revealed: the outcome is easily gamed, intentionally or unintentionally, by the person or team assessing the vulnerability.

Reinforcing its subjective nature, even NIST doesn't know how to define likelihood nonsubjectively:

"The term likelihood, as discussed in this guideline, is not likelihood in the strict sense of the term; rather, it is a likelihood score. Risk assessors do not define a likelihood function in the statistical sense. Instead, risk assessors assign a score (or likelihood assessment) based on available evidence, experience, and expert judgment. Combinations of factors such as targeting, intent, and capability thus can be used to produce a score representing the likelihood of threat initiation; combinations of factors such as capability and vulnerability severity can be used to produce a score representing the likelihood of adverse impacts, and combinations of these scores can be used to produce an overall likelihood score." [9]

Because there is no objective definition of likelihood, the results will not be consistent or even congruent between iterations, assessors, reviewers, or projects.

4.7.6.2   Other Rubrics

Now that we've made the difficulties about likelihood and related metrics clear, we can more easily assess the various rubrics available for their applicability to design-phase vulnerability scoring. First, let's look at those which are not suitable:

FDA's 2014 premarket guidance [10]

This rubric has some minor value, but only considers two equally weighted attributes of a vulnerability (exploitability and severity). The approach is overly simplistic and open to manipulation of the final score.

NIST's risk determination [11]

Heavy reliance on likelihood invalidates this rubric.

DREAD

Overly subjective; no longer supported.

CVSS 3.x [12]

This version of CVSS removed "collateral damage" (i.e., severity) from v2, the single most important factor to consider when scoring medical device vulnerabilities. It is possible to spoof this value by manipulating other attributes in this rubric, but then why use it if it is such as poor fit for medical devices?

NIST's CMSS [13] and CCSS [14]

These are very close variants of CVSS v2, but they introduce Likelihood in the form of general exploit level and perceived target value.

MITRE's medical device rubric [15]

A greatly expanded variant of CVSS v3 created by MITRE, this rubric was designed to be inclusive of multiple users' perspectives, including HDOs and MDMs. This variant introduces a large number of additional questions that must be answered to inform the attribute settings—45 questions in all—which must be assigned to each vulnerability before a base metric can be computed! Although it is a good tool for its intended purpose—evaluating deployed devices—this rubric is less suitable for design-phase vulnerability scoring because of this additional burden. In the initial assessment of a design, hundreds of vulnerabilities may be identified, and each would need to have 45 answers entered by the assessor before determining a base score.

The authors like this rubric as it is trying to address all the needs (this has never been attempted by any other rubric) for all the different domains where vulnerability and threat scoring needs to be performed. It was an ambitious effort, but basing it on CVSS v3 harmed it more than helped. When CVSS v4 is released it will be interesting to see if this rubric can or cannot be mapped into the CVSS v4 model.

IVSS [16]

This is an industrial variant of CVSS v3, and has several interesting additions to it, such as cascading consequences and process control consequences. However, at its base level, it is oriented toward a fully designed and released product: fundamental to its base scoring are attributes such as report confidence and exploit maturity.

OWASP risk rating [17]

This scoping rubric is very simple and is a member of the likelihood over impact scoring group. Since it relies on likelihood it is not suitable for design-phase vulnerability scoring.

EPSS [18]

Presented at the Black Hat 2019 conference, this rubric is a predictive variant of CVSS, where the probability of a vulnerability being exploited is considered by basing this on previous past exploits. Analysis of past common vulnerabilities and exposures (CVEs) is brought to bear to help quantify exploit probability (likelihood!) of each vulnerability. Unfortunately, the vast quantity of CVEs in the database are related to MIS/IT systems, not embedded devices, so many potential vulnerabilities are left unconsidered, such as those requiring physical attack vectors. Also, existing CVEs are skewed towards the simplest of attacks, such as buffer overflows, which causes more complex attacks to be ranked as (relatively) lower risks. Finally, because of restating likelihood as exploit probability, this rubric would not be suitable for design-phase vulnerability scoring.

Next, let's consider two rubrics that are suitable for design evaluations:

CVSS v2:

As we hope we've already demonstrated, this rubric is an example of a rubric that is simple to perform yet returns a final composite score based on appropriately weighted attributes. This rubric requires a quantity of 7 attributes to be assigned to each vulnerability before a base metric can be computed. Severity is supported in the form of collateral damage, which allows for flexibility in supporting impacts to patient safety and efficacy and well as business risks.

RSS-MD [19]

This rubric is a medical device variant of CVSS, requiring 11 attributes to be assigned to each vulnerability before a base metric can be computed. It is a good rubric to utilize for design-phase vulnerabilities with one minor exception—it has an attribute called scope of impact, which would not be relevant at the time of a medical device's design. Fortunately, the premature inclusion of this attribute can be easily corrected by setting it to "all" for each of the discovered vulnerabilities, which produces a minimal impact on the final calculated score. Because of its close relationship with CVSS, this rubric also shares some of its shortcomings, such as poorly defined attack vector settings. This rubric also focuses exclusively on patient safety impacts and is unable to account for other types of negative impacts, such as financial and reputational risks. Its patient focus also excludes the utilization of the medical device under attack as a pivot point into a network with a larger attack surface. However, even considering these limitations, this rubric should still be considered a viable candidate for scoring design time vulnerabilities.

4.7.6.3   Final Notes on Scoring Rubrics

Most of these rubrics were not developed specifically with application to medical devices, and none were designed to score design vulnerabilities during a secure development lifecycle. So, while it is possible to utilize CVSS v2 and RSS-MD for medical device design vulnerability scoring by slightly broadening one's interpretation of some of the scoring attributes, further standardization work needs to be performed before the medical device industry will have a formalized rubric well-suited for this purpose. This work includes: developing a rubric specifically for evaluating medical device designs; flexibility and proper weighting to account for vulnerabilities typical of embedded devices, such as portable and home use devices, rather than trying to stretch MIS/IT approaches to fit; and robust severity metrics that account both for risks to patient (safety and efficacy) as well as potential threats to the medical device manufacturer's business model (reputation and profitability).

4.7.7 Informal Approaches to Vulnerability Assessment

Although we are promoting STRIDE as our preferred formal method of decomposing a design into design vulnerabilities, there are other methods in use, such as informal decomposition of a design. These include whiteboarding sessions and threat trees. This section addresses some of these inferior approaches; reading it will not make you a better security engineer, but it might help you avoid a few pitfalls.

Try this mental exercise. Consider a development project example and imagine yourself in front of a whiteboard. Can you list all changes to be made to the software to prevent ransomware (a threat) from affecting your product? How do you know you have captured all of the necessary changes? Can you think of one more? A dozen more? When do you know you have a complete list?

Similarly, attack trees as commonly used are not helpful during product design. In addition to the reasons discussed under threat modeling, attack trees can also introduce a false sense of the attack surface. As an example, consider a simple use case of opening a locked door. (Figure 4.6)

(Note: for anyone unfamiliar with threat trees, the nodes with the arched undersides are OR nodes. The nodes with the flat bottoms are AND nodes. The boxes are all terminal leaf nodes.)

Is this a comprehensive threat model? It looks nice, and it seems that someone put some time into considering the possible attacks. But, again, because it is an informal approach to decomposition of a design, how do the leaf nodes lead to mitigations? Consider the burst through door leaf node—it leaves a lot of ambiguity as to the nature of the attack. Does it indicate the attacker is kicking the door in, using a battering ram, or charging through it with a tank? These have very different mitigations needed to protect against each separate type of attack.

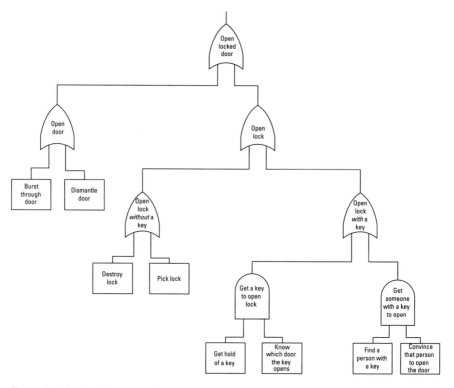

Figure 4.6   Locked-door attack tree.

Also, how would we know whether the model is complete? The attacker could block the latch from engaging (the technique used during the Watergate break-in), or socially engineer a situation where the door doesn't get locked in the first place, or take a photograph of the key and reproduce the key from the photograph, or attack the lock with foil impressioning. Some of these are obvious omissions (such as blocking the latch), but others (reproduction from a photograph, foil impressioning) are attacks that would probably only be anticipated by security experts.

Unfortunately, the majority of software engineers have had no training in security, so popular approaches which propose that informal threat modeling can consist of whiteboarding, brainstorming, threat trees, trust boundaries, and so on, end up being performed by people with little to no knowledge, experience, or training in security. This yields results that can cause extra work to be performed, or worse, work being performed in the wrong areas of a device while needed work goes overlooked.

To be clear, informal threat modeling is useful when employed by cybersecurity subject matter experts (SMEs) to clarify and document their thinking on a given area of security. But unless being used by such an expert, this tool doesn't assist a staff software engineer who is in the process of developing a medical device.

As with all the tools and techniques presented in this book, these represent the current state at the time of this book's writing. We anticipate that many, if not all, of these, will be deprecated in the coming years; threat trees are one example where this will likely occur sooner rather than later. The excellent work that was recently conducted in the EU by the TREsPASS team, which demonstrated unique methods of data visualization for large data sets and utilized attack tree linearization and attack clouds, holds promise for other tools that may build on their approach [20].

4.8 Development Lifecycle: Implementation Phase

Four major cybersecurity activities should be performed concurrently with implementation:

- Static application security testing (SAST);
- Fuzzing test of key communication mediums;
- Unit testing boundary analysis;
- Penetration testing.

Performing all of these tests concurrently with development ensures more accurate monitoring of project progress, easier implementation of fixes (authors have just completed working on the software being tested, so they are still familiar with the code), and more complete coverage of mitigations for discovered vulnerabilities.

SAST

Static analysis tools analyze software source code syntax and enforce secure coding standards. Various code security guidelines are publicly available, such as SEI CERT C and MISRA 2012 for a variety of computer languages. SAST should be performed often on the codebase to avoid a single huge effort needed to keep the codebase compliant. One of the best ways to accomplish this is to schedule it directly into the build cycle. Enforcing adherence to coding standards is good practice because it prevents poorly implemented software from becoming a vulnerability that may later be leveraged by an exploit. SAST is further discussed in Chapter 8.

Fuzzing

Fuzz testing is a negative software testing methodology that feeds a program, device, or operating system interface with malformed, unexpected input data in order to find defects. This malformed data can be randomly generated or can be based on experience and use data sets known to be problematic in a given scenario or for a given interface.

Fuzzing should be performed on all communication mediums implemented in the system. These communication mediums can be wired or wireless, and need not transport key or critical data to be suitable candidates for fuzz testing. Typical mediums to fuzz include Ethernet, Wi-Fi, Bluetooth low energy (BLE), USB, serial ports, proprietary low power RF, MedRadio, and inductive.

Fuzzing can take multiple forms, such as randomly altering packet structure and timing, or it can also attack the contents of the packet itself, such as the fields and values being received.

Fuzz testing should be performed for as long a period of time as is practical to the project. Keep in mind that time is the measurement of the success of a device under test, so many millions of iterations should be allowed during the testing regime.

Boundary analysis

Boundary value analysis is a black box type of testing, which stress-tests the supplied values (e.g., parameters) given to a function versus the intended range of values.

As an example, consider a function that expects a passed signed integer value of 8 bits to have a value of -100 to +100. As such there should be at least 6 defined test cases, exercising:

- −100 and +100;
- −101 and +99;
- −99 and +101.

Penetration testing

The goal of penetration testing is to get an independent indication of vulnerabilities that may exist in the system so that mitigating controls can be identified. Penetration testing is a negative black box form of testing where a security expert attacks the system in an unplanned and unstructured manner to identify any obvious vulnerabilities, as well as to provide an indication of the difficulty in exploiting said vulnerabilities. The amount of time needed by the security expert to exploit a vulnerability should be considered as valuable information in deciding if the vulnerability requires mitigation.

Penetration testing should not be left until late in the development lifecycle, as the testing can take significant time to perform, and the results of testing (detected vulnerabilities) become increasingly expensive to mitigate late into development. Instead, penetration testing should be performed concurrently with implementation activities, scoped to specifically defined aspects of the system as they are implemented.

As an example, the Ethernet-based command and control structure of the device may have reached the status of feature complete, while the rest of the device is still under development. Therefore, penetration testing efforts would only attempt to compromise this Ethernet interface, ignoring the other parts of the device. This results in multiple penetration tests being performed on separate sections of the device. Obviously, planning and scheduling are an important part of this type of testing.

4.9 Development Lifecycle: Verification and Validation Phase

In a properly structured and executed secure development process, few security-related activities are performed during V&V testing.

The only cybersecurity-related V&V testing should be for:

1. Design vulnerabilities discovered during the vulnerability analysis;

2. Impacted safety and efficacy;

3. Subsequent confirmation by the risk management process.

Even if mitigations were implemented for these vulnerabilities, there should still be V&V testing coverage to ensure the safety of the device.

4.10 Development Lifecycle: Release Phase/Transfer to Production

Congratulations! Your new product has passed all the hurdles of development, including all the security-related activities and regulatory approval. You now need to move the software elements into manufacturing.

The first cybersecurity-related hurdle to this transition is: did your build system get compromised and subsequently infect your software deliverable? Are you about to manufacture and ship thousands of devices pre-infected with malware?

The correct approach to preventing this is to utilize a tightly controlled build system maintained exclusively for this purpose. Preferably, this build system is offline and passes all infrastructure security checks, including virus detection, Host IDS, Network IDS, and IPS. The software deliverables should all be signed or have hashtag values for each of the deliverables.

4.10.1 Three Different Transfer Models

At this point the options diverge into three possible approaches:

1. Have nonvolatile components commissioned and programmed with the deliverable at a value-added reseller (VAR) who has implemented secure handling of components.

2. Have the deliverables transferred into a secure programming environment in your company's internal manufacturing facility. There, deliverable integrity shall be confirmed on each programming iteration as well, and endpoint protection tools installed on these programming systems will block attempts to exfiltrate the deliverables out of the programming tools as would be performed during an insider attack. (Competitors may attempt to pay your employees to steal your intellectual property, therefore, freedom of movement for your software deliverables must be restricted).

3. Have the deliverables transferred into a secure programming environment in a contract manufacturing (CM) facility. There, deliverable integrity shall be confirmed on each programming iteration as well, and endpoint protection tools installed on these programming systems will block attempts to exfiltrate the deliverables out of the programming tools. A certificate management system should also be put into place to allow remote issuance of certificates to be programmed into the devices. This management system is supported by auditing activities against the CM. As an example, if the certificate manager is commanded by the MDM to issue 1,000 certificates and when auditing is performed 950 units where shipped, there had better be 50 units awaiting destruction (or rework) by the auditing team. This prevents the CM from creating off-the-books devices for their own profit.

Further manufacturing and production considerations are discussed in Chapter 5.

4.11   Development Lifecycle: Sales Phase

The medical device industry is highly regulated and compliance-driven. This has led to a generally conservative culture, not interested in taking on too much business risk or stepping too far ahead of their competitors. The shift toward openness on cybersecurity, including supporting the sharing of cybersecurity information with customers, has been no different in that regard.

Even the leading luminaries of medical device manufacturing would never publish marketing material stating that a medical device is completely secure, with no risk of hacking! And for good reason. They would instantly be on every hacker and security researcher's target list, and their devices would be subjected to intense security scrutiny (by security researchers) and attacks (by malicious actors). Additionally, in part because it wasn't advisable to advertise security features in order to improve sales, MDMs have historically viewed these features as offering little value to the company, leading to attempts to reduce financial impact by minimizing security activities.

Yet, changing economic conditions, regulatory guidances, and customer demand have brought about a new opportunity. Security-related deliverables allow MDMs to differentiate themselves from their competitors' security postures, without risking the aforementioned blowback from making public claims, and establish reputations as industry leaders. Cybersecurity now offers a competitive advantage, and thus is becoming easier to implement in a more favorable and motivated business environment—and is increasingly appreciated by customers.

This ongoing change is also influencing the sales process. It used to be (and still may be in some cases) that sales organizations were skilled in deflecting any questions or requests on the topic of cybersecurity. That is now evolving into sales becoming a cooperative partner in the security discussion by proactively positioning the value of security to their customers, answering specific questions, and responding to requests (e.g., in the RFI or RFP process). Sales' responsibility now includes support of the presales process by providing upfront security documentation, negotiating security-related terms as part of the contract, and acting as a trusted partner in case of a security escalation (whether resulting from the device not meeting requirements, an actual incident, or something else). This development has fundamentally changed the sales phase and has made it a much more integral part of the secure development lifecycle. For a deeper discussion of sales' role in cybersecurity, see Chapter 5.

4.12 Development Lifecycle: End of Life Phase

For MDMs, the generalized steps under the umbrella concept of end of life (EOL) are end of production, end of sales, and end of support. Possible causes for each stage include business considerations, competitor products, market shifts, or component obsolescence or unavailability. There is no regulatory or standards framework that dictates details about these steps, and they may vary based on manufacturer or device type, but in general the underlying considerations with regards to cybersecurity are similar. All of these best practices are subject to alteration by contractual relationships between the MDM and the HDO as well as local regulations in the region where the device is marketed.

End of production does not end the MDM's security responsibilities as devices still in use (or devices warehoused by the manufacturer or distributor) may still need to be managed and maintained from a security perspective and the MDM will still need to continue their postmarket surveillance.

Depending on the type of device, end of sale may follow sooner or later after end of production. Although this step does have impact on customers and device users (as replacement devices or accessories my no longer be available), it, again, does not necessarily end the MDM's security responsibilities.

End of support is the only step where there may be impact on the MDM's cybersecurity responsibilities since this is where the continued provision of security mitigation (e.g., patches, updates) is likely to end. MDMs must notify their customers well in advance of their intended end of support (EOS) date; three years' notice is typically considered a reasonable minimum.

However, we should recognize that when it comes to cybersecurity, EOS may not be clearly defined or easy to determine. For example, an MDM may continue to sell and support a device that contains a component that is no longer supported by its OEM (e.g., Windows 7). In this case, the device as a whole is still being supported by the MDM, but a security-critical component is not. This creates a challenging situation for the HDO and other customers (e.g., home healthcare agencies), and may present the MDM with insurmountable cybersecurity exposures.

The MDM EOL as part of the development lifecycle typically is related to an entire product line or type. It is distinctly different than the HDO EOL (also referred to as decommissioning), which is specific to the individual physical device. For example, an MDM may end the life of an entire line of infusion pump as they move on to the next generation, whereas an HDO may retire one of its pumps due to age but keep the remainder of them operating.

For the MDM, there are two general touchpoints with the HDO EOL: (a) MDM will need to design features into the device to support HDO EOL (e.g., data scrubbing); and(b) a given device (or batch) may have an expiration date (e.g., due to expected component life, like the battery). By law (e.g., US 21 CFR part 803), the MDM is required to calculate and specify the EOL date for a given device batch based on its date of manufacture and communicate it to their customers, as well as manage their own warehoused inventory accordingly.

For further discussion of the cybersecurity challenges arising from the MDM's device EOL versus the HDO's decommissioning, see Chapter 5.

References

[1]   TrapX Labs, "Anatomy of an Attack: MedJack (Medical Device Hijack)," May 7, 2015. https://trapx.com/trapx-labs-report-anatomy-of-attack-medical-device-hijack-medjack/.

[2]   Brunau, C., "Common Types of Ransomware," *Datto*, Aug. 22, 2019, https://www.datto.com/blog/common-types-of-ransomware.

[3]   Center for Internet Security, "Security Primer—Ransomware," 2020, https://www.cisecurity.org/white-papers/security-primer-ransomware/.

[4]   Whitwam, R., "Australian Police Warn of Ransomware USB Drives Showing Up in Mailboxes," *ExtremeTech*, September 22, 2016, https://www.extremetech.com/computing/236157-australian-police-warn-of-ransomware-usb-drives-showing-up-in-mailboxes

[5]   Bode, K., "Researchers Find 'Anonymized' Data Is Even Less Anonymous Than We Thought," *Vice*, Feb. 3, 2020, https://www.vice.com/en_us/article/dygy8k/researchers-find-anonymized-data-is-even-less-anonymous-than-we-thought.

Ehrenkranz, M., "Researchers Reveal that Anonymized Data Is Easy to Reverse Engineer," *Gizmodo*, Jul. 23, 2019, https://gizmodo.com/researchers-reveal-that-anonymized-data-is-easy-to-reve-1836629166.

Campbell-Dollaghan, K., "Sorry, Your Data Can Still be Identified Even if It's Anonymized," *Fast Company*, Dec. 10, 2018, https://www.fastcompany.com/90278465/sorry-your-data-can-still-be-identified-even-its-anonymized.

[6]   Thompson, S., and C. Warzel, "How to Track President Trump," *The New York Times*, Dec. 20, 2019, https://www.nytimes.com/interactive/2019/12/20/opinion/location-data-national-security.html.

[7]   https://www.first.org/cvss/v2/guide.

[8]   Gates, C., "Vulnerability Scoring: Comparing CVSS V2 & V3 for Medical Devices Pt. 1," *Velentium*, Dec. 19, 2019, https://www.velentium.com/blog/vulnerability-scoring-comparing-cvss-v2-to-v3-for-medical-devices.

[9]   NIST Joint Task Force Transformation Initiative, "SP 800-30r1: Guide for Conducting Risk Assessments," September, 2012, https://csrc.nist.gov/publications/detail/sp/800-30/rev-1/final.

[10] "Content of Premarket Submissions for Management of Cybersecurity in Medical Devices; Guidance for Industry and Food and Drug Administration Staff," *FDA Guidance Documents*, Oct. 2, 2014, https://www.fda.gov/regulatory-information/search-fda-guidance-documents/content-premarket-submissions-management-cybersecurity-medical-devices-0.

[11] NIST Joint Task Force Transformation Initiative, "SP 800-30r1: Guide for Conducting Risk Assessments," September, 2012, p. 81, https://csrc.nist.gov/publications/detail/sp/800-30/rev-1/final.

[12] FIRST CVSS Special Interest Group, "Common Vulnerability Scoring System v3.1 Specification Document," June, 2019, https://www.first.org/cvss/v3.1/specification-document.

[13] LeMay, E., K. Scarfone, and P. Mell, "The Common Misuse Scoring System (CMSS): Metrics for Software Feature Misuse Vulnerabilities," July, 2012, https://csrc.nist.gov/publications/detail/nistir/7864/final.

[14] Scarfone, K., and P. Mell, "The Common Configuration Scoring System (CCSS): Metrics for Software Security Configuration Vulnerabilities," December, 2010, https://csrc.nist.gov/publications/detail/nistir/7502/final.

[15] Chase, M., and S. Christey Coley, "Rubric for Applying CVSS to Medical Devices," September 2019, https://www.mitre.org/publications/technical-papers/rubric-for-applying-cvss-to-medical-devices.

[16] Bodungen, C., "Industrial Vulnerability Scoring System (IVSS)," https://securingics.com/IVSS/IVSS.html.

[17]   The OWASP Foundation, "The OWASP Risk Assessment Framework," https://www. owasp.org/index.php/OWASP_Risk_Rating_Methodology.

[18]   Jacobs, J., S. Romanosky, B. Edwards, M. Roytman, and I. Adjerid, "Exploit Prediction Scoring System (EPSS)," August 31, 2019, https://arxiv.org/abs/1908.04856.

[19]   QED Secure Solutions, "Risk Scoring System for Medical Devices (RSS-MD)," https:// www.riskscoringsystem.com/medical/techspecmedical.pdf.

[20]   https://visualisation.trespass-project.eu/.

5

Security in Production and Sales

5.1 Production

Transferring design to and maintaining security during production is a critical task (see Chapter 4). Additional considerations need to be made to secure products during the production process, during postmarket servicing, and rework on returned devices. During the design phase of the development lifecycle, the needs of the production line requirements must be considered, such as the anticipated volume of units that will be built each year, the number of lines to be supported, the geophysical location of the lines, the time required to manufacture each unit (takt time), the communication medium being utilized to connect to devices during the production process, and the number of software applications needed on the device (testing routines utilized in production in addition to final production code), to name a few.

Further considerations related to securing the manufacturing process include:

- If you are uploading new software images to a unit being built and the communications medium is slow (such as BLE), do you have the time to perform each upload while maintaining secure processes?

- Will unique identifiers be written to each device, or is the identifier based on another variant value, such as the serial number? How is device software versioning (down to a security-relevant level) managed?
- Does the device include cryptographic security technology and if so, how will unique symmetric key or unique asymmetric key pairs be written to the device?
- How will these cryptographic values be managed? Will they be local or hosted? How will they be managed across possible multiple production locations?
- The use of a backend server means there has to be an IP connection to the production line, which incurs an additional risk of hacking a production line workstation, manufacturing equipment, or even devices in production.
- Is there some out-of-band (OOB) medium being utilized, such as a 2D barcode that needs to be printed and associated with the correct unit being manufactured, and if so, how is that matched with the device's security-relevant software version?
- Are the programmable components being fully programmed by a so-called value-added reseller (VAR) before they reach the production line? Are they completely blank components? Are they being programmed by a VAR with just the bootload and software update agent, for loading on the production line?
- At what step in the production line is the JTAG (debugging port) being disabled?
- At the end of the line, are the units being placed into a low power mode for storage? Is there a risk that they could be awoken by an attacker to deplete the battery life while in storage?
- If there is just one software image loaded into the programmable components, which includes the production line's testing routines, how are these routines permanently disabled at the end of the production line?
- If separate software is installed for testing or calibrations functions, how is it removed or its function disabled at the end of the production line?

These types of considerations should be evaluated jointly by the cybersecurity design support and the production teams. Joint evaluation should cover the range from assuring secure manufacturing infrastructure and processes, secure engineering-to-manufacturing transfer, and security features designed into the device to support these activities. The production team will be in a better position to evaluate the production impact of cybersecurity controls and design mitigations being proposed to be performed by, or on behalf of, production. In situations where personnel from the specific production organization should not be involved in a joint review—for example, because the main production will be handled by a contract manufacturer who should not be exposed to the design process—the MDM should involve an in-house or independent production engineer in the cybersecurity planning process, followed by ongoing audits of the contract manufacturer to ensure that all security processes are being performed and documented.

5.1.1 Production Line Functionality Left Enabled in a Shipped Device

One of the most common cybersecurity mistakes MDMs make is to leave production line functionality enabled in a shipped medical device. This has resulted in many attacks that have been documented by security researchers, and in several cases, this has led to a great deal of bad publicity for the MDM. In at least one instance, a major MDM (who shall remain nameless) compounded its mistake of this type by allowing a senior-level executive to immediately respond to the news with a broad denial of any security issues. In order to reinforce its case and credibility, the security researcher then responded by making public more details about how to exploit this particular vulnerability. Doubling down on bad decisions, the MDM responded to this latest detailed disclosure with the same senior-level executive riffing about how this latest disclosure of detailed vulnerabilities simply wasn't true. By the time of the third response, more sensible heads prevailed and the MDM started addressing the issue in a responsible way, as opposed to a knee-jerk denial reaction. (See Chapter 10 for more on this topic).

What was really at work in this incident are three different potential failures of the production process.

The first failure was leaving the production functionality (e.g., for testing purposes) enabled in a shipped device. The production line does have support needs for nonstandard operating conditions. These needs can include calibrating sensor inputs and driving the device to extreme limits of delivery or mechanical position with the removal of all alarms and safeties. This type of unrestricted functionality, while necessary in a production environment, can have a devastating effect upon a patient and is often trivial to exploit in a fielded device. Ease of exploit is frequently compounded by a total lack of authentication for accessing production-line functionality—either because nobody even considered implementing authentication, or because the additional time requirement for authenticating to each device during production was deemed too costly.

The bottom line is as simple as this: If any functionality is not related to the normal essential performance of the medical device, then that functionality must be disabled before shipping.

Fortunately, mitigating this issue is actually quite straightforward, and there are multiple ways to address it. One option is developing dedicated production software that's loaded into the device exclusively for certain tests and is overwritten upon successful completion of them. However, reprogramming the device on the production line may not be desirable due to the negative effect on the throughput of the production line as reprogramming a device can be a time-consuming operation.

Another option is to include production-line functionality in the final fielded firmware, but to disabled it once that functionality is no longer needed. At this point, a production line station would issue a command that permanently disables the production-related functionality. Note that to be effective, this must be a one-way command with no facility to reverse the disabling of the production-related functionality.

As with the previous option, the production functionality would be included in the final fielded firmware and remain enabled, however, this time with cryptographically significant authentication and authorization required before any functionality can be activated.

The second process failure in our example was the lack of consideration and direct support for the production line in the design of the device. All production-related and reworking functionality needs to be considered and directly supported in the initial design of the device, not as an afterthought. A common example of this bad practice is leaving the JTAG port enabled in a fielded device.

The third process failure in our example related to the vulnerability intake process and is more of a decision-making and communication issue, as evidenced by the lack of established procedure for responding to the initial disclosure made by the security researcher. The lifecycle of this type of security event should be planned out in advance, including the content of the text to release to the public, and should not be left for some senior executive to decide on the fly what should and what shouldn't be said to the public.

Consideration should likewise be made in advance to establishing friendly contacts in news organizations that can be counted upon to accurately represent your preplanned press release, as opposed to hyperbole that may increase sales of their publication. The preplanned press release text should be a multidisciplinary effort including representatives from customer-facing communications, legal, production, regulatory affairs, and engineering. It should also be clear who is authorized to communicate to external entities (press, regulators, law enforcement) and what part of the communication they are responsible for (e.g., business, clinical, technical). In this way, the MDM can have an intelligent, logical, well-planned, and well-reviewed approach to presenting the vulnerability to the public (including customers) in a timely and informative fashion and avoid knee-jerk reactions that can severely damage the MDM's credibility. The communication responsibilities, best practices, and preparations for vulnerability management are similar to those discussed for incident response in Chapter 11.

5.1.2 Factory Service and Rework

Some thought should also be paid to the remanufacturing and factory-level service of the medical device. The design will need to support such functionality as:

- Uniquely secured (i.e., not a universal password) functionality to download logs from a device (as JTAG is disabled).
- Uniquely secured functionality to return a device to factory default values and erase all PII and PHI which is stored in the medical device.
- Uniquely secured functionality to return a device to a low power mode for storage.
- Uniquely secured functionality to return a device to either the production line test functionality or a separate service level test functionality.
- Utilizing firmware updating functionality for devices being serviced (see Chapter 6). The methodology described is also applicable to the factory, with some adjustments or specifically designed secure functionality to accomplish the update.
- Uniquely secured functionality to place a device into a configuration mode, such as supplying Wi-Fi credentials when a medical device is being installed in a given working environment.

Again, it is recommended that the MDM involve a production engineer during design so that his or her experience can be used to help design the secure pathways that will be needed for service and rework.

5.1.3 Securing Production Infrastructure

Another critical area that requires attention is the security posture of the production environment itself. Responsibility for assuring protection of the production environment typically lies with the enterprise security team, or a dedicated operational security team. From a cybersecurity perspective, there are several objectives to be met:

- Protect the production environment from cybersecurity compromise to avoid costly shutdown and prevent recalls or regulatory notifications;

- Assure integrity of the finished product so that a) only approved configurations are shipped, and b) the product itself is not compromised by malware or backdoors that could be exploited by an adversary;

- Assure correct and timely management and phase-in of product changes that occur in the normal lifecycle of a product or may occur in response to and as part of the mitigation of a newly discovered vulnerability;

- Protect products from security compromise through the hardware or software supply chain.

Of course, all of these are related and require a holistic approach to assure production security and integrity are met and maintained (Figure 5.1). If even one of the above is absent, the finished product is at heightened risk of a costly cyber compromise. A cybersecurity risk assessment and management approach specific to the production environment is advisable.

There are, of course, differences depending on the type of product. A hardware medical device with embedded software is different from a pure software medical device (SAMD). Furthermore, although not a topic of this book, production security and integrity considerations should also be applied to non-software products in the biotech or pharma industries as a production line security compromise could result in shutdown or product quality impact. For example, in 2017 the NotPetya ransomware outbreak at a leading pharmaceutical manufacturer resulted in steep financial losses as well as shortages of drugs and vaccines and to make matters worse, may not be covered by cyber insurance [1].

As in all areas of cybersecurity, a mature approach requires that we address people, process, and technology. Production staff needs to be properly trained

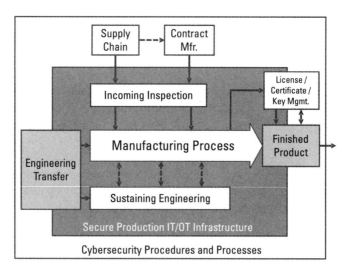

Figure 5.1 Securing production infrastructure.

on not only specific secure-build processes, but also on general cybersecurity hygiene and best practices. They should not, for example, use a production line PC to check their personal email over lunch (which also, as discussed later, should be prevented by technical means); neither should they bring in a USB data carrier with their family photos to use as a wallpaper or screen saver.

Key to assuring that staff can execute in a way that does not compromise security are properly defined and assigned roles and responsibilities. Who is responsible for which areas of security (e.g., deploying and managing security technology, installing new production equipment, performing data transfer in and out of the production environment, and so on)? Sustaining engineering has a key responsibility in addressing security-related issues and tasks as they arise during the production process.

Based on individuals' roles and responsibilities, regularly scheduled training refreshers as well as updates on emergent security topics of relevance should be provided and documented. Furthermore, staff needs to have the ability (and confidence) to detect potential cyber risks and compromise and take appropriate action.

From a process perspective, the areas that need to be addressed are:

- Receiving engineering transfer of new product or updates.

- Processes to secure license/key/certificate lifecycle management and deployment during production (discussed in more detail in Chapter 10). Insufficient lifecycle management processes or security of the license/key/certificate infrastructure can result in devices with compromised security (e.g., when using cryptography-based security technologies) or financial losses (e.g., compromised licenses).

- Managing change during the production process (e.g., phasing in updates as a result of a newly discovered vulnerability) leading to engineering change and requiring mitigation before the finished product ships. Note: for the purpose of product change management during production, a manufacturer may define a point of no return after which changes will no longer be implemented and will require a field upgrade. This will, obviously, depend on the type of product, production process variables, and the type of change.

- Production V&V (engineering V&V asks "Did we design the right product?", whereas production V&V asks "Are we building the right product?"). Includes test runs and audits in cooperation with customers (to test the entire chain including shipment, delivery, installation, documentation, and license/key deployment).

- Production environment security policies and procedures (how to keep the production environment secure).

- Specific policies and procedures addressing staff onboarding, termination, and training (as well as supportive processes like incident response, etc.).

From a technology and infrastructure perspective, we need to assume that the production environment is a mix of information technology (IT: workstations, servers, storage) and operational technology (OT: SCADA and ICS devices, robots) integrated on a dedicated production network. Serendipitously, many of the security challenges in the OT environment are similar to that of the medical device ecosystem itself. For example, we often find an older infrastructure that, from a security perspective, is difficult to maintain and protect. A well-planned and well-executed defense in depth approach is the answer, assuring security across all control points, complemented by network architecture and network-based security to compensate for the remaining vulnerabilities that cannot be addressed by commercial security technology. Applicable solutions are:

- Endpoint security for traditional IT devices in the production environment (workstations, servers, smart devices). However, due to the complexity of today's endpoint security products, they are typically not suitable for nontraditional endpoints with limited system resources and the need for highly predictable behavior.

- Host intrusion detection and prevention systems (HIDS/HIPS) as an alternative security approach for IT endpoints (more complex to implement but lower on system resources and resulting in a more secure posture) as well as OT/IoT endpoints (may require cooperation with the manufacturer).

- Network-based security technologies ranging from traditional firewalls, (network) intrusion detection and prevention system ((N)IDS/(N)IPS), security gateways, and anomaly detection solutions can all be applied to improve the overall security posture of an enterprise at the network level.

- Data loss prevention (DLP) to monitor and prevent loss of confidential data like source code or other intellectual property.

- Security management, including security infrastructure management console, security operations, and incident detection, and response capabilities.

- Other security technologies should be applied as appropriate, for example virtual private networks (VPN) and multifactor authentication (MFA) for remote access as well as, on an as-needed basis, data-at-rest and data-in-motion encryption.

In today's threat environment of targeted, sophisticated, and stealthy attacks conducted by highly capable cybercriminals and adversarial nation states, none of the above by itself will provide sufficient security. The combination and cooperation of most if not all of the suggested technologies is needed.

Furthermore, in addition to the above technologies the production environment requires a secure network architecture that is properly segmented (using traditional VLAN or more recent software defined network (SDN) approaches) and minimizes integration points (which in turn minimizes security exposure) with other enterprise IT systems. For example, the production network may need to have integration with other enterprise systems (although, some may only need to be temporary) for engineering-to-production transfer, or enterprise resource planning/management (ERP/M) systems like scheduling, inventory, materials, and order management. However, unnecessary integrations should be avoided, for example between the production network and general business systems, like email. And certainly, a production network should not be connected to a coffeemaker [2].

Some additional areas that typically require attention, yet may be overlooked for the production environment, are:

- Incident response, typically including the key steps of planning, practice, identification, triage, containment, eradication, recovery, and follow-up. Good forensic data collection is critical to, for example, understand the extent and duration of a compromise and to determine if the last 24 hours of production were affected, or the last week, or longer. Another important aspect of incident response is the communication with and responsibilities of stakeholders external to the production organization: legal, regulatory, executive management, and if warranted, customers and users. For more details, see Chapter 11.

- Physical security of the production environment to prevent malicious and nonmalicious tampering with infrastructure and products. For example, the last thing you want is that an eager engineer enters the production environment with a critical software update or test tool on a USB stick.

- Which leads to the next topic, data carrier handling. Data carriers, like USB thumb drives are extremely useful but are also known to transfer malware into even air-gapped or heavily isolated networks. Data carriers may be used for internal purposes (e.g., software or documentation transfer) or external purposes (e.g., maintenance of production equipment). Ideally, any production environment should be able to function without requiring sneaker nets, but there may be practical limitations and exceptions. For that purpose, strict procedures should be in place to prevent the contamination of the production environment. This should include strict handling requirements, hardware limitations, and pre-scanning of data carriers on stand-alone scan stations.

• Integrity and security maintenance of the production environment requires timely deployment of updates and patches and, as needed, cooperation with the vendor of the equipment in use (SCADA, ICS, robots).

The above describes the most generic production environment as an example. Obviously, special considerations apply depending on the specific situation, which could include a geographically distributed production infrastructure, contract manufactures and supply chain of off-the-shelf components, as well as access requirements for suppliers, vendors, or other partners.

It is essential to properly manage any third party that interacts with the production environment (e.g., a supplier of hardware or software components). As part of the incoming inspection, checks of security critical properties should be applied (e.g., assessment of component version and compatibility, or performing malware scans). Components should have the ability to be inspected for their integrity (e.g., through the use of checksums or hashes for software).

In addition, suppliers and contract manufacturers should be contractually required to comply with established security requirements and procedures, and compliance should be confirmed by audit.

Unfortunately, the authors are aware of several cases where manufacturers have been compromised by their supply chain or where manufacturers delivered devices to HDOs that had malware installed on them out of the factory. This highlights the importance for a secure and well managed production environment to assure integrity and quality of the finished product.

5.2   Security Considerations in the Sales Process

Traditionally, MDMs have been very conservative compared to other industries. The nature of medical devices and their regulatory environments have (mostly) discouraged MDMs from publicly making unsupportable claims to achieve sales and from pursuing risky product development activities. This conservative approach has also extended into the field of product security claims. Some MDMs have been ahead of the industry in coordinated disclosures, where disclosed vulnerabilities are presented, triaged, and hopefully mitigated. However, it has been uncommon for an MDM to make public claims about the positive aspects of their product security for fear of becoming a hacking target.

This reticence in bragging about following good security practices means that product security has not historically been part of any sales efforts. However, changing conditions have opened the doors for a market-savvy MDM to leverage product security as a sales advantage in direct communications with potential customers—without making public or unsupportable claims which might single them out as a target.

As a result of increasing security and safety concerns, HDOs are now including security requirements in their purchasing contracts and typically expect that the MDM supplies three separate deliverable artifacts. These are:

- Description of device software composition: Software Bill of Materials (SBOM);

- Description of device security properties: MDS^2 [3];

- Security commitments (technical and process) in contracts between the HDO and the MDM.

Some HDOs, based on guidance provided by the FDA, may stipulate delivery of additional artifacts that help them assess a device's security posture. These may include:

- Documentation of hazard (risk) analysis, mitigations, and design considerations (including list of cybersecurity risks considered and cybersecurity controls applied);

- Traceability matrix that links cybersecurity controls to the identified risks;

- A plan for postmarket management of software updates and patches;

- Documentation of software integrity controls (i.e., assurance that the software remains free of compromise);

- Instructions and specifications related to cybersecurity controls (e.g., antimalware software or firewall management) as well as secure deployment, handling, and integration;

- Evidence assuring software security (e.g., evidence of testing or certification).

Both human and machine-readable SBOMs are a useful sales tool as they indicate to an HDO that the MDM is taking their cybersecurity responsibility seriously and performing best practices in the total product lifecycle. As HDO sesurity stances mature, the availability of machine-readable SBOMs for use in the HDO's asset management systems will serve as a huge incentive for acquiring SBOM-based medical devices. Given the relative ease with which an MDM can create and manage SBOMs, this will be well worth the MDM's effort. For more information about SBOMs, please reference Chapter 8.

5.2.1 MDS²

The Manufacturer Disclosure Statement for Medical Device Security (MDS²) was initially developed in 2008 through the cooperation of National Electrical Manufacturers Association (NEMA) and Health Information and Management Systems Society (HIMSS) Medical Device Security Task Force, in collaboration with multiple industry associations, government agencies, and other stakeholders. It provides a basic overview of the device security in 41 questions, allowing medical device manufacturers to describe to their customers the basic security and privacy properties of a specific medical device. This includes things like the operating system and version, type of network connection, the ability of the operator to install antivirus software, or what and how much protected health information (PHI) is stored on the device and whether it is transient or permanent.

Simply put, the goal of the MDS² is to assist the HDO's professionals who are responsible for security-risk assessment in the management of medical device security issues. The MDS² can also become an important sales tool as it allows an HDO to make a security-informed decision about its purchase.

Filling out the MDS² is purely voluntary and typically needs to be specifically requested by the purchasing entity. However, a few manufacturers have made it their practice to publicly disclose the MDS² by linking to it on their web page.

A major update was released in 2013 (81 questions) and in October 2019 a new revision of the MDS2 was released by NEMA. The scope of the MDS² has increased dramatically, now supporting 216 questions!

The new MDS² has 23 sections that now include:

- The device's specifications;
- Automatic logoff;
- Management of PII;
- Remote access;
- Data integrity controls;
- Audit controls;
- Personal authentication;
- Authorization controls;
- System hardening;
- Data backup and disaster recovery;
- Emergency access;
- Antimalware;
- Supported connectivity;

- Node authentication;
- Security guidance;
- Software upgrades;
- Availability of an SBOM.

This new MDS² improves the granularity of the information captured including compatibility information that may facilitate communication between medical devices. The MDS² does not specify service processes, support commitments, or other activities that are better addressed via contracts (see Section 5.3).

For the moment, it is unclear how many HDOs are actually utilizing MDS² for any activity in the hospital, including procurement or security event handling. However, given its obvious utility in a variety of roles at the HDO, we anticipate that it will be leveraged more thoroughly in the future for post-procurement activities and increasingly adopted as a requirement for procurement evaluations.

In the meantime, the MDS² provides a strong sales tool for MDMs wishing to highlight their secure development practices and preparedness for post-market cybersecurity support. As HDOs increasingly seek more security information from manufacturers, it may help differentiate your medical device from your competitors'.

5.3 Cybersecurity in Contracts

While contracts have been and will continue to be the vehicle that defines and constrains the business relationship between the HDO and the MDM, detailing such items as price, service, and warranty, there are now additional elements that are more commonly being added to contracts by the HDOs in the form of cybersecurity assurances and ongoing services to be provided by the MDM.

These new terms are being utilized to define and formalize the cybersecurity posture between the HDO and the MDM. These efforts are far from having mandated standards, but some of the publicly available examples [4] and voluntary standards and guidelines [5] do paint an overall picture of where the industry is heading.

Some common threads run between most of these artifacts, and by reviewing these common threads we can start to level-set some of our expectations for the language we are going to see in new contracts.

The following are paraphrased topics that are now commonly found in contracts between the HDO and the MDM and include technical as well as process requirements:

- Secure development lifecycle:

- The MDM commits to following a current best-practices secure development program;
- There is documentation to substantiate this secure lifecycle;
- The MDM follows industry best practices for internal staff training and awareness;
- The MDM agrees to auditing by third-party experts;
- Conformance and compliance with existing laws and regulatory guidelines.
- Security requirements:
  - Fallback modes of operation to safely maintain essential clinical performance during and after an attack;
  - Features to allow restoration of device functionality and configuration data.
- Security updates and patches:
  - Setting up specific time periods (measured in days) for the MDM to notify the HDO of any discovered vulnerabilities;
  - Setting up specific time periods (measured in weeks) for the MDM to provide a patch to mitigate the vulnerability; the exact time period may depend on vulnerability risk level, the higher the risk the sooner the update should be provided;
  - All patches will be cryptographically authenticated (i.e., signed or hashed);
  - Updates to TPSC (antimalware, operating system, etc.) patched within a specific time period (measured in weeks).
- Security events at the HDO:
  - Point of contact at the MDM;
  - Support for identification and remediation efforts.
- Security events at the MDM:
  - Notify the HDO within a specific time period (measured in weeks);
  - Respond to the events in a specific period of time (measured in days);
  - Support the investigation of security issues and active incidents.
- Liability allocation and remedies for events:
  - MDM shall maintain cybersecurity insurance;
  - MDM is liable for costs incurred by the HDO for nonconformance related events:
    - Retesting the assets attached to the HDO's network;
    - Loss of productivity;
    - Applying the mitigation patch;

- Liability of MDM for all related damages.
- SBOM:
  - MDMs agree to provide an up-to-date and accurate TPSC bill of materials;
  - TPSCs are traced to any CVEs that are reported in vulnerability databases, such as NVD;
  - SBOM is routinely assessed for any newly disclosed vulnerabilities.
- Encryption:
  - All communications between devices are to be encrypted;
  - All data at rest is to be encrypted.
- Authentication:
  - User to device;
  - Device to network;
  - Device to device/backend;
  - Login credentials shall use a long, complex, and unique password;
  - Support two-factor authentication, especially for a high privilege or remote access.
- Logging:
  - Create log entries for authentication;
  - Create log entries for data access;
  - Create log entries for security events and incidents.
- Rights for the HDO to conduct penetration testing:
- Rights for public disclosure of a discovered vulnerability under responsible disclosure.

Contracts are very flexible and some of the language we currently see being suggested for use in contracts will need to be revised, either to be more realistic (e.g., "the device will have no vulnerabilities" is extremely unrealistic!), or to specify clearly which party is going to shoulder which ongoing costs of meeting the obligations set forth in these contracts.

5.4 Managing End of Life

End of life (EOL) is a very subjective concept. When examined from the HDO's perspective, it can mean:

- Decommissioning the device to waste disposal (landfill) or electronic recycling;
- Ownership transfer of the device to another HDO or back to a vendor;
- Return of a leased or loaned device.

From the perspective of the MDM, EOL can mean:

- Ownership transfer of the device's product line to another MDM;
- End of the device's warranty period;
- End of the device's support period;
- End of life for a component or TPSC utilized in the device necessitates redesign of the device;
- End of the device being marketed;
- End of the device being manufactured;
- Estimated end of life;
- Hard-specified end of life (such as implantables). See C.F.R. 21 part 803.3 [6].

For MDMs, the generalized steps under the EOL umbrella are end of production, end of sales, and end of support. MDMs must communicate any of these events to the customers well in advance of the end of support date; typically, 3 years is considered a reasonable minimum period of notification. From a cybersecurity perspective this results in three major challenges:

1. Following end of support from the MDM, there will be no further support or vulnerability patching, so risk-based decisions have to be made to determine whether continued use of the medical device by the HDO is warranted.

2. For any of the HDO's types of EOL, a common theme emerges. Before the medical device leaves the hands of the HDO, any data being stored in the device needs to be erased. This includes PHI, PII, and any configuration settings, including security-related data, such as device or user credentials.

3. Following a transfer of ownership of the medical device to a new HDO, the MDM may not be able to locate the new owner to notify them of discovered vulnerabilities or to provide security patches to the device.

These three problems have three possible solutions, with varying degrees of effectiveness:

1. The MDM should, along with the end of support notification, also provide technical descriptions as to how the device could be placed into a more secure configuration or environment, such as disabling communication mediums, segmenting the network that is connected to the device, and so on. The MDM should take this as an opportunity to support the HDO by offering newer updated device options

in an attempt to maintain the business relationship with an existing customer.

2. The restore to factory default functionality should be a mandatory feature in any device that stores or buffers any patient data. This functionality should be gated by some form of authentication to prevent patient data from being destroyed in a normally functioning system that is not at EOL.

3. Unfortunately, it can be very difficult for the MDM to track geographical location changes and new owners of a medical device to inform them of cybersecurity events and software upgrades. This is one more reason why MDMs need a product security portal on their web presence where customers can come to them to seek out security information about their newly acquired used medical device.

References

[1] Griffin, R., K. Chiglinsky, and D. Voreacos, "Was It an Act of War? That's Merck Cyber Attack's $1.3 Billion Insurance Question," *Bloomberg*, Dec. 3, 2019, https://www.insurancejournal.com/news/national/2019/12/03/550039.htm.

[2] Sullivan, T., "How a Coffee Machine Brewed Up Ransomware, and Other Startling Findings in the HIMSS Cybersecurity Report," *Health Care IT News*, Aug. 1, 2017, https://www.healthcareitnews.com/news/how-coffee-machine-brewed-ransomware-and-other-startling-findings-himss-cybersecurity-report.

[3] NEMA, "Manufacturer Disclosure Statement for Medical Device Security," Oct. 8, 2019, https://www.nema.org/Standards/Pages/Manufacturer-Disclosure-Statement-for-Medical-Device-Security.aspx.

[4] Indiana University Health, "Vendor Relations," 2020, https://iuhealth.org/about-our-system/vendor-relations.

[5] Health Sector Coordinating Council Joint Cybersecurity Working Group, "Health Industry Cybersecurity Supply Chain Risk Management Guide," Oct., 2019, https://healthsectorcouncil.org/wp-content/uploads/2019/10/Health-Industry-Cybersecurity-Supply-Chain-Risk-Management-Guide-v1_Final_PDF.pdf.

Health Sector Coordinating Council Joint Cybersecurity Working Group, "Medical Device and Health IT Joint Security Plan," Jan., 2019, https://healthsectorcouncil.org/wp-content/uploads/2019/01/HSCC-MEDTECH-JSP-v1.pdf.

Loughlin, S., "In Contracts with Device Vendors, Mayo Clinic Emphasizes Security," *Biomedical Instrumentation & Technology*, Jan./Feb. 2016, p. 53, https://aami-bit.org/doi/pdf/10.2345/0899-8205-50.1.53.

[6] 21CFR803, "Medical Device Reporting," Apr. 1, 2019, https://www.accessdata.fda.gov/scripts/cdrh/cfdocs/cfcfr/CFRSearch.cfm?CFRPart=803&showFR=1&subpartNode=21:8.0.1.1.3.1.

6

Medical Device Manufacturer Postmarket Lifecycle

6.1 Understanding FDA Expectations

Medical devices were put under the oversight of the FDA when congress enacted the Medical Device Amendments of 1976. At the time of signing, President Gerald Ford stated: "The Medical Device Amendments of 1976 eliminate the deficiencies that accorded FDA 'horse and buggy' authority to deal with 'laser age' problems," observing that, "When well designed, well made, and properly used, [medical and therapeutic devices] support and lengthen life. If poorly designed, poorly made, and improperly used, they can threaten and impair it [1]."

In 1976, President Ford would have had no awareness of cybersecurity or the risks it would someday pose to medical devices, yet his statement applies equally well to today's cybersecurity risks. It remains a testimony to the flexibility of that amendment and the actions resulting from it that today's FDA is able to protect America's population by setting expectations for medical device security. International regulators in many countries, as discussed in Chapter 2, have since taken similar action.

Following the Cybersecurity Premarket Guidance of 2014, the next step taken by the FDA was in December of 2016 when it released the Postmarket Management of Cybersecurity in Medical Devices Guidance. This guidance addresses the back-end of what we now call the total product lifecycle (while the

Content of Premarket Submissions for Management of Cybersecurity in Medical Devices addresses front-end activities). The Postmarket Guidance states that MDMs should take a risk-based approach to cybersecurity throughout the device's entire lifecycle (Figure 6.1), including:

- Monitoring of cybersecurity signals;
- Maintenance of device cybersecurity posture;
- Identification of potential vulnerabilities and resulting risks (severity and exploitability);
- A mechanism to provide timely mitigation (e.g., security patches) to devices in the field [2].

Another of the FDA's expectations for postmarket activities is the sharing of vulnerability and other cyber risk information within the larger medical device and health delivery community. Yet, as of this writing some four years after its release, these required postmarket activities still largely come as a surprise to many MDMs. This is not to say that it hasn't had some impact for the better: since the publication of the Postmarket Guidance, vulnerability disclosure has increased more than 5-fold [3].

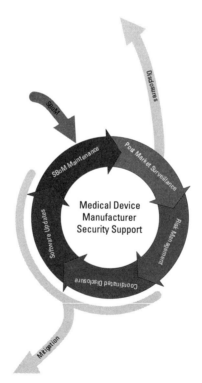

Figure 6.1   MDM security support lifecycle.

These MDM activities are time-consuming and stand outside of traditional project-oriented activities. Moreover, the level of security and software expertise required to perform them also puts these activities outside the skill set of the traditional sustaining engineer. This has contributed to difficulties and challenges in introducing these postmarket activities into many MDM's existing processes.

Another area of contention is the time period called out in the postmarket guidance for fielding a security patch to a vulnerable device. Specifically:

"As soon as possible but no later than 60 days after learning of the vulnerability, the manufacturer fixes the vulnerability, validates the change, and distributes the deployable fix to its customers and user community such that the residual risk is brought down to an acceptable level" [4].

This is a laudable security goal set forth by the FDA, but this time period does not align with the reality of operations as normally performed by an MDM. While the FDA has graciously stepped out of being involved in fielding a new security-only patch, and thus removed some delay in getting this out to the field, there still exists the legal liability of creating new patient risks in any rushed fielding of new patches with abbreviated testing regimens. Reperforming validation testing of the device with the new patch could in and of itself take longer than 60 days. A quick glance at the history of Microsoft's patches to its operating systems will illustrate the frequency of introducing new unintended and undetected issues when fixing the original issue. MDMs and the patients they serve simply cannot afford that level of failure in medical devices. While a final determination has not yet been made regarding the exact timetable for deploying new mitigations, it is generally agreed that MDMs must act promptly to secure their devices against postmarket risks. This chapter addresses those postmarket activities and advises MDMs on how to monitor and securely update their devices already in the field.

6.2 Postmarket Surveillance and Related Activities

The process of postmarket observation for security introduces a new set of asynchronous activities that are not initiated by any other MDM development or support activity. These postmarket surveillance activities include:

- Continuously monitoring for disclosed vulnerabilities related to third-party software components (TPSC) utilized in your products;
- Continuously monitoring for vulnerabilities in the MDM's application software;
- Practicing coordinated vulnerability disclosure (vulnerability intake, analysis, and mitigation) for TPSCs and MDM application software.

- Some surveillance activities go hand-in-hand with the need to engage with the market and communicate any findings and recommended actions to the MDM's customer base;

- Engaging with and supporting product users and consumers regarding product security;

- Maintaining active involvement in an ISAO to be informed of current threats.

6.2.1   Monitoring TPSC Vulnerabilities

This process is linked to the SBOM activity, as the SBOM serves as the TPSC content authority for any version of the software in a device. On a continuous basis, the SBOM's contents are reviewed against disclosed vulnerabilities and threats that may impact the operation of your device, or other devices in the device's installed environment (such as an attack utilizing your device as the entry point into an HDO's network).

The sources of these disclosed vulnerabilities can vary significantly, including the National Vulnerability Database (NVD) [5] (which usually has significant delays in posting new vulnerabilities), Vulners [6], a Github or other software version control system's repository history, ISAO disclosure, or popular media. Monitoring some of these sources can be automated, but mostly this is a manual process.

Once any vulnerability is discovered (either externally or internally to the organization), it needs to be evaluated for its impact on the medical device's operation. It should be noted that it is very common for a TPSC with a vulnerability to be utilized in a device, yet have absolutely no impact on the security standing of that device. This situation can exist either because the TPSC's vulnerable element is not functionally utilized by the device, or because the device design renders it unable to be exploited.

As an example, OPENSSL v1.0.1f (the last version susceptible to the Heartbleed vulnerability) has 494 defined macros that control an untold number of the preprocessor's conditional compilation functionality (i.e., "#IF-DEF"). This functionality delimits blocks of program statements that are compiled only if the specified condition is true (see Figure 6.2). These directives are frequently nested and can contain combinations of multiple macro names.

The result is a very large number of possible binary executables options that may or may not include the Heartbleed vulnerability from this one version of OPENSSL—depending upon the status of 494 macros. It is entirely possible, even likely, that the application software utilizing the OPENSSL library does not call any of the vulnerable functions, which may render an exploit based on them much more difficult to execute.

```
#ifndef NO_FORK
        int multi=0;
#endif

#ifndef TIMES
        usertime=-1;
#endif

        apps_startup();
        memset(results, 0, sizeof(results));
#ifndef OPENSSL_NO_DSA
        memset(dsa_key,0,sizeof(dsa_key));
#endif
#ifndef OPENSSL_NO_ECDSA
        for (i=0; i<EC_NUM; i++) ecdsa[i] = NULL;
#endif
#ifndef OPENSSL_NO_ECDH
        for (i=0; i<EC_NUM; i++)
                {
                ecdh_a[i] = NULL;
                ecdh_b[i] = NULL;
                }
#endif


        if (bio_err == NULL)
                if ((bio_err=BIO_new(BIO_s_file())) != NULL)
                        BIO_set_fp(bio_err,stderr,BIO_NOCLOSE|BIO_FP_TEXT);

        if (!load_config(bio_err, NULL))
                goto end;

#ifndef OPENSSL_NO_RSA
        memset(rsa_key,0,sizeof(rsa_key));
        for (i=0; i<RSA_NUM; i++)
                rsa_key[i]=NULL;
#endif
```

Figure 6.2   Example from OPENSSL.

This kind of variability in the resultant binary output means each software component with a disclosed vulnerability needs to be evaluated in consideration of how, specifically, that software component is being utilized in the final product before any definitive statement can be expressed about the impact of the vulnerability on the device.

However, every disclosed vulnerability needs this type of investigation into its impact, and if it is a risk that can be exploited, then mitigations need to be explored. Mitigation options include:

- A field security patch that includes a new version of the third-party software component that has been mitigated by the supplier;
- A temporary workaround that is communicated to the consumer of the medical device;
- A medical device recall.

Once an exploitable vulnerability is discovered, the MDM should perform a coordinated disclosure, which exposes the status of this investigation and response process to the ISAO and directly to the end-user. Depending upon the nature of the user (HDO, home healthcare, international consumer, etc.) the methods for keeping them informed about the risk status will vary.

6.2.2   Coordinated Vulnerability Disclosures

No matter how security-focused an MDM's development lifecycle is, vulnerabilities will exist in the released device at some point in the future (that could be tomorrow or in five years). This can be caused by a number of factors:

- A flawed process within the secure development lifecycle;
- A missed implementation bug;
- An intentional backdoor/vulnerability created by an MDM insider during development;
- Unintended configurations or communications;
- A previously unknown vulnerability in a TPSC (including cryptographic libraries);
- Advances in computing rendering existing cryptographic primitives useless;
- A completely novel or unanticipated type of attack (i.e. AI-assisted side-channel analysis) [7].

So it has to be assumed there will be vulnerabilities discovered in the device after it has been released to the field. These vulnerabilities are discovered through a variety of ways, including accidental operation, focused analysis, reverse engineering, intentional testing, including subjecting the product to a variety of stress tests, and side-channel analysis.

The person performing this discovery process is referred to as the reporter. The reporter can be anyone, from a casual user, security researcher, hacker, or even someone internal to the MDM.

When a new vulnerability is discovered, there are a number of paths this process can follow:

- *Full disclosure:* Where the reporter publishes the analysis of the vulnerability as quickly as possible, making the analysis information accessible to everyone without delay or restriction.

- *Responsible disclosure*: Where the reporter informs the MDM of the vulnerability first, sharing the analysis with nobody besides the MDM until either:
  - The MDM releases mitigation to address the vulnerability (e.g. remediation via a patch) or initiates a voluntary recall (if warranted).
  - An agreed-upon period of time has elapsed. Note that this period of time should tolerate the added burden placed on MDMs versus nonmedical reporting scenarios. This additional burden can include many months of testing prior to being able to release new patched software. The primary interest here should be the safety of the patients, not an arbitrary period of time.

- *Nondisclosure*: Where the vulnerability information is not shared, and this bifurcates into:
  - Disclose analysis only to the MDM, typically under contract, bug bounty, or NDA;
  - Sell to the intelligence community to remain in place as a secret vulnerability to be used in a future attack (AKA a Zero-Day Attack).

- *Coordinated vulnerability disclosure (CVD):* Where the reporter informs the coordinator and either the coordinator or the reporter and coordinator together informs the MDM. Limited information is released to the public until such time as the vendor has provided a mitigation (e.g., remediation via an approved patch). This has the added benefit of the coordinator being an independent judge who can monitor the MDM's progress in the investigation, mitigation, and distribution of a patch. The coordinator can encourage both the reporter and the MDM to act responsibly in addressing the vulnerability in a timely fashion. This also facilitates more effective CVD and discourages its misuse as a mere PR option rather than a legitimate risk mitigation tool.

These methods of disclosure have been the source of conflicts for decades between vendor and reporter. MDMs often fear that any vulnerability disclosure would be harmful to their products and thus their business model, while the reporter often feels that MDMs are not addressing their security concerns in a timely fashion, and either misrepresenting the serious nature of the vulnerability or just ignoring the vulnerability claim completely until the reporter has no other alternative than to publicly disclose [9]. In some cases, it appears that both sides are at fault [10]. In other cases, no party appears to be at fault, but a series of poor judgments and miscommunications are performed by both parties, to the detriment of all.

Fortunately, this war between reporters and MDMs seems to be cooling down (while other institutions still seem rooted in the past [11]), in large part due to the increasing adoption of coordinated vulnerability disclosure practices, which is viewed by both sides as a compromise position beneficial to both the reporter and vendor and is being increasingly recognized and recommended by regulators around the world.

There are many excellent standards for implementing coordinated vulnerability disclosure (NTIA, ISO, CERT, MDIC, IETF, and Google's Project Zero) as well as good organizations (CERT/CC, DHS CISA, H-ISAC) that will assist the researcher or the MDM in publicizing the vulnerability.

MDMs may find CVDs uncomfortable to perform, feeling that they are tarnishing their good name and ruining their position in the marketplace. Several years ago, when cybersecurity awareness was low and the realities of the threat landscape widely misunderstood, that was a valid business concern; though it should be noted that it was never a valid concern from the point of view of protecting patients and end-users. To avoid public exposure, many MDMs would conduct a silent recall and some MDMs may still be reluctant to do otherwise, despite FDA efforts to incentivize responsible management of the disclosure. However, many MDMs have taken the lead and implemented excellent disclosure sites, as illustrated in Table 6.1.

Coordinated disclosure has changed from being a negative PR event to a positive one, where an MDM can celebrate their proactivity in addressing security issues in medical devices before patients are harmed. In today's environment, disclosures help maintain and develop public trust in the MDM brand,

Table 6.1
Examples of Coordinated Vulnerability Disclosure Sites

Company	Disclosure Site
Abbott	https://www.abbott.com/policies/cybersecurity/cybersecurity-coordinated-product-disclosure.html
BD	https://www.bd.com/en-us/support/product-security-and-privacy
Boston Scientific	https://www.bostonscientific.com/en-US/customer-service/product-security.html
Medtronic	https://global.medtronic.com/xg-en/product-security.html
Phillips	https://www.philips.com/a-w/security/coordinated-vulnerability-disclosure.html
Roche	https://diagnostics.roche.com/global/en/legal/vulnerability-and-incident-handling-policy.html
Siemens	https://new.siemens.com/global/en/products/services/cert.html#SecurityPublications
Stryker	https://www.stryker.com/content/stryker/us/en/about/governance/cyber-security.html

which can lead to increased market share for its current and future devices, as well as limiting the MDM's exposure to the perception of negligence.

6.2.3 Engagement with End-Users

In the last decade, we have seen an increasing amount of MDM engagement with end-users (both HDOs and home health care), and this is especially true when it comes to supporting cybersecurity. While some of these postmarket supporting activities and communication will be driven by contract language between the MDM and the HDO, in other cases it is as simple as addressing the security of your products on a webpage and having a portal where end-users and security researchers can contact the MDM to report product security vulnerabilities, voice concerns, and ask for assurances about the security of the medical device.

This communications portal should at a minimum take the form of an email address in one or all of the following formats:

- security@[domain name];
- security-alert@[domain name];
- secalert@[domain name].

All emails sent to any one of these addresses should have an auto-responder to inform the sender that the email was received. Following that, a human should respond in a timely manner (24–48 hours) to address the questions and concerns of the sender, including informing them, if appropriate, that the reported vulnerability is being escalated to a team that will investigate the claim.

Many MDMs also make an effort to assure the confidentiality of the communication by publishing a public encryption key that can be used to secure the message. This is often seen as beneficial to both the reporter and the MDM.

As the vulnerability is progressing through the investigation there should be regular and timely emails to the reporter, informing them of the progress throughout the investigation and, if deemed necessary, mitigation development and deployment. This type of continual engagement with the reporter will help the MDM in case any further information or details of the vulnerability is needed from the reporter during the investigation, as well as inform the reporter that the MDM is taking their claim seriously, which should also discourage the reporter from publicly disclosing the vulnerability prematurely or in an otherwise uncontrolled manner.

All communications should be in a respectful and thankful tone, no matter how incredible or fanciful the claim being made. In one case we are aware of, the sender reported that "their medical device infected their cell phone following a large flash of light emanating from their medical device!" Cybersecurity is a scary topic for most people. Combine that fear with medical devices and the level of anxiety of the average layperson can be severe.

Maintaining positive public relations during the postmarket period depends on reporters feeling that their concerns are heard and are being taken seriously. A reporter's concern does not need to be technologically valid to threaten the MDM's business model. If the MDM responds rudely or dismissively, its response may provoke the reporter to broadcast the concern and/or response, which can result in a backlash of fear and/or negative sentiment toward the MDM, either of which can damage the brand.

6.2.4   ISAO

An information sharing and analysis organization is an organization set up to coordinate cybersecurity information sharing in industries that are considered part of a nation's critical infrastructure. The health care delivery and medical device industry are part of that group.

Full participation in an ISAO is strongly recommended by the FDA. ISAO participation must be documented if the MDM wants the FDA to credit their participation. Criteria for active participation are as follows:

1. The manufacturer is a member of an ISAO that shares vulnerabilities and threats that impact medical devices.

2. The ISAO has documented policies pertaining to participant agreements, business processes, operating procedures, and privacy protections.

3. The manufacturer shares vulnerability information with the ISAO, including any customer communications pertaining to cybersecurity vulnerabilities.

4. The manufacturer has documented processes for assessing and responding to vulnerability and threat intelligence information received from the ISAO. This information should be traceable to medical device risk assessments, countermeasure solutions, and mitigations [14].

For maximum efficacy, the information must be shared in as close to real-time as possible. Participation includes identifying standards and guidelines for robust information sharing and analysis related to cybersecurity risks, incidents, and best practices.

The FDA has signed Memoranda of Understanding with the Department of Homeland Security (DHS), H-ISAC, MedISAO, and Sensato to facilitate information sharing about potential or confirmed medical device cybersecurity vulnerabilities and threats. To motivate MDMs to join an ISAO, the FDA has lowered its expectations for when the MDM's recall must be reported for a security correction, as shown in Figure 6.3. For additional information, see Product Recalls.

Expected key outcomes for belonging to an ISAO include:

- Improved understanding of vulnerabilities in medical devices;
- Improved solution development for the entire stakeholder community;
- Harmonized best practices for device security information sharing;
- Improved efficiency to market while simultaneously improving security, safety, and privacy profiles for devices and associated networks.

Clearly, there are many benefits to belonging beyond meeting regulatory expectations. To summarize, it is in any MDM's best interest to belong to an ISAO!

6.3 Remotley Updating Devices in the Field

"Updating the software in fielded medical devices is easy—just look at all the examples from semiconductor manufacturers, or rely on existing Linux package

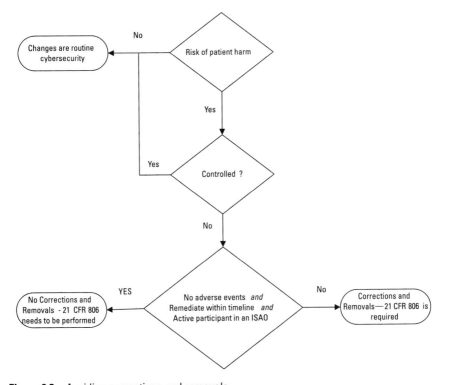

Figure 6.3 Avoiding corrections and removals.

managers to get the job done!" This, or something similar, has been the uninformed opinion of nearly every engineer when they first start down the path of updating device software in the field. Most realize rapidly that nothing could be farther from the truth! Creating a secure, reliable, robust system to provide update functionality is, in itself, a major project requiring skills in a whole host of different technologies. This functionality is referred to by several different names, including firmware over the air (FOTA) and device firmware update (DFU).

MDMs are required by the FDA to be able to provide field updates in a timely manner to address security issues as they are discovered. Even if you are following the best practices for securing a device—even if you have completely secured the device to the maximum degree possible at the time of development—the tools and techniques of attackers are constantly evolving and improving. What may be secure today can be insecure tomorrow. For this reason alone, having the capability to update devices in the field is mandatory.

However, field updates create, at a minimum, four potential vulnerabilities:

1. Exposure of intellectual property or security mitigations by reverse-engineering the update image deliverable;

2. Corruption or modification of the update image deliverable;

3. Rollback of device software to a previous update version with exploitable vulnerabilities;

4. Use of the update process as part of a supply chain attack.

Mitigations to these implicit vulnerabilities must be included in every field updateable system.

There are many examples of cybersecurity update failures that have occurred in just about every industry, including automotive [16], personal computers [17], smartphones [18], printers [20], and just about everything else [21]. In some instances, updates have been used to intentionally brick the device (i.e., render it completely unable to function) [22]. The need for a controlled, well-engineered update process cannot be overstated.

Starting at the top and going down to individual devices, we begin with an update engine. This is the database of the update images, and is responsible for:

• Securely storing the update images.

• Determining the correct update image (if any) for a given model number. This includes determining the device's current hardware and software version, as well as verifying the upgradability of a given device.

• Building the secured deliverable package, which may contain one or more update images and update metadata.

- Reporting on the version status of any device.
- Reporting on the status of a given update rollout, including depth of progress in the rollout.
- Communicating the update deliverable package to the device/system to be updated, in a device-agnostic manner.
- Logging all actions performed, including additions, deletion of update images, associations between different compatible versions of update images, execution of update rollouts, and so on.
- Extensive levels of user authentication and authorization to perform actions, such as multiparty approvals, with multifactor level authentication for each party.

As seen in Figure 6.4, there are a multitude of paths and communication mediums that could be involved in the process. Communication infrastructure along these pathways should be designed to be agnostic about the nature of what is being transmitted to avoid creating unnecessary roadblocks in update delivery and reporting.

In Figure 6.4 it should be noted that although there are a large number of potential communication mediums, ultimately the update deliverable package arrives at the medical device(s), where each medical device may have several internal components that also may or may not need to be updated, plus several other medical devices or device accessories that may need updating. This results

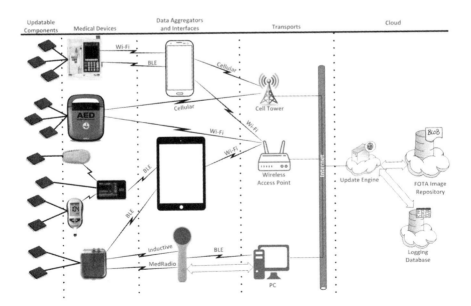

Figure 6.4   Potential paths/mediums involved in field software updates.

in a whole constellation of elements that need to be updated as part of the same update procedure. Coordinating this process, especially in a way that allows for errors that may require aborting the update and rolling back to the previous version, is a difficult and complex task.

To avoid a whole class of failures, including those represented in the failure examples previously noted, the target element that is being updated should have sufficient nonvolatile program memory storage to accommodate two complete and separate images of the executable (note: this is not just buffering in random access memory (RAM), this needs to be a fully executable image in the nonvolatile location in which it is placed). By this approach, two separate versions of the executable can reside in the device and allow for rollback to the previous version if a failure occurs in any part of the update process (Figure 6.5).

Once a program has complied and the resultant output file is created, it has to be converted into an update image file via an image maker program that adds:

- Versioning information;

- Target hardware information;

- Salt appended to the beginning and end of the compiled output file (this is necessary to avoid encryption of common compiler output formats and patterns that could be utilized to attack the encryption. Salt is a nonsecret value used in a cryptographic process to ensure that the results of computations for one instance cannot be reused by an attacker);

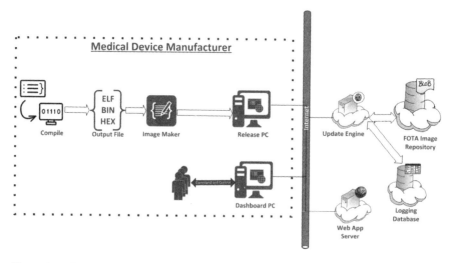

Figure 6.5　Update package assembly and delivery.

- Encryption;
- Signature or hashtag.

This data structure is now the update image file. It is added into the FOTA image repository and traced to any other update image files with which it may need to interact. As communication protocols and content changes, incompatibilities between update targets can occur if one target is updated and another target is left with an older incompatible version of its software.

When an update must be performed, the process can be enabled via a dashboard portal into the update engine, which allows specific hardware to be targeted for updates and specifies the update images to be pushed to those targets. Before the process can begin, multiple people in the organization should be required to be authenticated to enable the update process. This is recommended to avoid an insider attack where unintended firmware images could be pushed to all of the devices in the field. Best practices should be utilized in this authentication process, such as 2FA/MFA.

A controlled series of predetermined steps should be followed to perform a device update. Here's an example:

1. Communications is established with the target device.

2. The target device is notified of the presence of a newer version.

3. Depending on the device type and use case, the target device may or may not require a user physically onsite with the device to approve the update process. This is due to the wide variety of medical device types, such as life support devices that cannot be interrupted in any manner while in active use and so must be approved by on-site personnel for safety, versus small sensor devices that have no user interface on the device at all. The nature of the end-user involvement is defined and controlled by the MDM via the dashboard portal as part of the update process.

4. The target device reports back all of the hardware configurations and all of the software versions currently in use in its system to the cloud update engine. This may include multiple CPUs per device and multiple devices and accessories, depending on the system.

5. Using the versioning information for each system and the tracing information provided when each update image was added to the FOTA repository, the cloud update engine builds a secured deliverable package (which may consist of multiple compatible update images). This secure deliverable package is uniquely encrypted for each device and instance of the process, as well as signed or key-hashed to ensure the integrity and authenticity of its origin.

6. This secured deliverable package is then communicated back along the path of communications. Data in motion security mitigations are not assumed or required to be in place, due to the secure nature of the deliverable package.

7. From this step forward, any errors or detection of compromised data is a reason to terminate the entire process and perform rollbacks to the previous version software if necessary.

8. Upon receipt of the secured deliverable package by the medical device, the integrity and origin of the package is confirmed.

9. The package is decrypted and the constituent update image files are communicated to the correct component or device in the system.

10. As each target component receives its update image, the integrity of the image is confirmed and then decrypted.

11. The appended salt is removed and discarded.

12. The version information is compared to the current version. The update image version must be greater than the current version for the update to proceed.

13. The plaintext version of the update image is written to its intended operating location in program memory.

14. Any preupdate activity that needs to be performed would be performed now, including any backout state to be utilized for recovery, if needed.

15. Each target device and component would confirm the integrity of the stored update image. This does not necessarily have to be a cryptographic operation and could be as simple as a cyclic redundancy check (CRC).

16. Utilizing whatever communication mediums are in place between these components and devices, each target indicates its integrity and readiness to proceed.

17. If any components do not report that they are ready to proceed, the entire update process on all devices and all components are halted. The failure report—or failure to report, whichever is applicable—is communicated back to the cloud update engine for logging.

18. If all components are ready, then the cloud update engine is informed that the update process was a success (and the success report is logged).

19. If all components indicate their readiness, then each component performs a reboot and allows the boot manager code to determine the most recent version of intact software to transfer execution to.

20. In the update image for each component, there may have been post-update activities to be performed (such as cleaning up any temporary data, converting persistent data structures' contents or indexing addresses, etc.). If present, these activities are now performed.

21. The system has successfully completed the update process.

The success or failure of each update attempt in the field is logged for reporting activities performed at the MDM and may be needed for review by the FDA or the DHS to confirm the progress of the update process against the number of devices in the field. This attrition rate should be balanced against patient risk and if the rate of updating the software is found to be too slow [23] or too problematic [24], additional methods may have to be pursued, such as additional lines of communication to the consumers of the medical device, and even the potential for a recall.

In all cases, the safety of the patient is paramount and must be the primary consideration when structuring the update process. A good example of this would be an implantable pacemaker where a vulnerability has been discovered, but no evidence of active or past exploits has been identified. This situation would require a much longer update period, since it would need the patient to go into the physician's office to have the update performed where emergency care could be provided if needed. However, if this same vulnerability were discovered in a case where devices were being actively exploited and patients were dying, the situation might justify an update being pushed to the patients' devices even without the support of emergency care capability immediately on-hand. This is just one example of the difficult decisions that need to be performed when updating a medical device in the field.

6.4 Product Recalls

In the United States, a recall is a method of removing or correcting devices that are in violation of FDA risk regulations. Failure to correct a violation could lead to legal enforcement.

There are three types of recall:

1. Voluntary recall under 21 CFR 7;

2. Mandatory recall under 21 CFR 810;

3. Corrections and removals under 21 CFR 806.

The vast majority of recalls are voluntary and are initiated by the MDM. Technically, a recall is "a firm's voluntary removal or correction of a marketed product that the FDA considers to be in violation of the FD&C Act and against which FDA would initiate legal action" [25].

All recalls are classified according to the relative degree of health hazard posed by the continued use of the medical device as follows:

I.   A situation in which there is a reasonable probability that the use of, or exposure to, a violative product will cause serious adverse health consequences or death;

II. A situation in which use of, or exposure to, a violative product may cause temporary or medically reversible adverse health consequences or where the probability of serious adverse health consequences is remote;

III. A situation in which use of, or exposure to, a violative product is not likely to cause adverse health consequences [26].

Recalls are one of the more frightening events that can occur for an MDM. Voluntary or mandatory recalls are both bad for customers, public relations, future regulatory activities, and the business. Recalls cost MDMs an average estimated $2 million dollars per recall, with costs for a single incident as high as $600 million, not including associated drops in the MDM's stock price and market share losses that potentially push into double-digit percentages [27]. Recalls can also cause stagnation in development as engineering resources are diverted to fix the problems rather than create new products, thus allowing competitors opportunity to gain on the MDM's market share. Clearly, recalls are bad, and in many cases can be avoided by implementing a robust software updating system. Furthermore, as previously explained, correction and removal recall reporting is affected by membership in an ISAO.

The vast majority of medical device recalls are not cybersecurity-related. More often than not, cybersecurity issues lead to a vulnerability alert and in severe cases may even trigger a warning letter or safety communication. However, during the 2019 calendar year, there were 16 such actions [28].

The best way to avoid having to perform a mandatory or voluntary cybersecurity-related recall is to design new and future medical devices with secure field-update capability so that vulnerabilities can be mitigated without exercising that option.

References

[1] Ford, G., "Statement on Signing the Medical Device Amendments of 1976," May 28, 1976, https://www.presidency.ucsb.edu/documents/statement- signing-the-medical-device-amendments-1976.

[2] "Postmarket Management of Cybersecurity in Medical Devices; Guidance for Industry and Food and Drug Administration Staff," *FDA Guidance Documents*, Dec. 28,

2016, https://www.fda.gov/regulatory-information/search-fda-guidance-documents/postmarket-management-cybersecurity-medical-devices.

[3] MedCrypt, "What Medical Device Venders Can Learn from Past Cybersecurity Vulnerability Disclosures," Jan. 2020, https://medcrypt.com/resources/ICS-CERT_Whitepaper 2020 pdf.

[4] "Postmarket Management of Cybersecurity…," Dec. 28, 2016, p. 22.

[5] https://nvd.nist.gov/.

[6] https://vulners.com/search?query=order:published.

[7] Bursztein, E., and J. M. Picod, "A Hacker Guide To Deep Learning Based Side Channel Attacks," *Defcon* 27, Las Vegas, NV, Aug. 8–11, 2019, https://elie.net/talk/a-hackerguide-to-deep-learning-based-side-channel-attacks/.

[8] Donohue, B., "Critical Security Vulnerabilities in Drug Injection Pumps," *Kaspersky Daily*, May 12, 2015, https://usa.kaspersky.com/blog/drug-pump-security-bugs/5331/.

[9] Kan, M., "Medical Device Security Disclosure Ignites an Ethics Firestorm," Computerworld, Aug. 29, 2016, https://www.computerworld.com/article/3113385/medical-device-security-ignites-an-ethics-firestorm.html.

[10] Krebs, B., "Iowa Prosecutors Drop Charges Against Men Hired to Test Their Security," *KrebsOnSecurity*, Jan. 31, 2020, https://krebsonsecurity.com/2020/01/iowa-prosecutors-drop-charges-against-men-hired-to-test-their-security/.

Wakabayashi, D., "He Won Praise for Halting a Global Cyberattack. Then He Was Arrested," *The New York Times*, Aug. 3, 2017, https://www.nytimes.com/2017/08/03/technology/cybersecurity-researcher-hailed-as-hero-is-accused-of-creating-malware.html.

[11] "Postmarket Management of Cybersecurity…," Dec. 28, 2016, p. 26.

[12] Greenberg, A., "Tesla Responds to Chinese Hack With a Major Security Upgrade," *Wired*, Sep. 27, 2016, https://www.wired.com/2016/09/tesla-responds-chinese-hack-major-security-upgrade/.

Howard, B., "It Won't Be the Last Time: Over-the-Air Glitch Bricks Lexus Infotainment," *ExtremeTech*, June 14, 2016, https://www.extremetech.com/extreme/229877-it-wont-be-the-last-time-over-the-air-glitch-bricks-lexus-infotainment.

[13] Morris, L., "A Dell Update Bricked My Computer," *Fstoppers*, Mar. 2, 2019, https://fstoppers.com/originals/dell-update-bricked-my-computer-344764.

Lovejoy, B., "Limited Reports of Catalina Installation Bricking Some Macs Via EFI Firmware", *9TO5Mac*, Oct. 24, 2019, https://9to5mac.com/2019/10/24/efi-firmware/.

[14] MJ, A., "3 Ways to Fix iOS 12/12.3/13 Update Bricked My iPhone," *Wondershare*, Dec. 26, 2019, https://drfone.wondershare.com/ios-12/update-bricked-iphone.html.

[15] Cui, A., M. Costello, and S. Stolfo, "When Firmware Modifications Attack: A Case Study of Embedded Exploitation," *NDSS Symposium 2013*, San Diego, CA, Apr. 23, 2013, http://ids.cs.columbia.edu/sites/default/files/ndss-2013.pdf.

[16] Brignall, M., "Samsung TV Owners Furious After Software Update Leaves Sets Unusable," *The Guardian*, Aug. 24, 2017, https://www.theguardian.com/technology/2017/aug/24/samsung-tv-buyers-furious-after-software-update-leaves-sets-unusable.

Palmer, A., "Not-So-Smart Sneakers: Nike 'Working to Fix' Software Update that Bricks its $350 Self-Tying Trainers, Leaving Them Unable to Tighten," *The Daily Mail*, Feb. 21, 2019, https://www.dailymail.co.uk/sciencetech/article-6731599/Not-smart-sneakers-Software-update-bricks-Nikes-350-self-tying-trainers.html.

[17]   Kieler, A., "Samsung Software Update Will Deliberately 'Brick' Remaining Galaxy Note 7 Phones," *Consumer Reports*, Dec. 9, 2016, https://www.consumerreports.org/consumerist/ samsung-software-update-will-deliberately-brick-remaining-galaxy-note-7-phones/.

[18]   Saxon, L., N. Varma, L. Epstein, L. Ganz, and A. Epstein, "Factors Influencing the Decision to Proceed to Firmware Upgrades to Implanted Pacemakers for Cybersecurity Risk Mitigation," *Circulation 2018*, Vol. 138, Sep. 18, 2018, pp. 1274–1276, https:// www.ahajournals.org/doi/pdf/10.1161/CIRCULATIONAHA.118.034781.

[19]   Lee, J., M. Henrich, P. Bibby, S. Mulpuru, P. Friedman, Y. M. Cha, and K. Srivathsan, "Pacemaker Firmware Update and Interrogation Malfunction," *Heart Rhythm Case Reports*, Vol. 5, No. 4, April 2019, pp. 213–216, https://www.heartrhythmcasereports. com/article/S2214-0271(19)30001-6/pdf.

[20]   "Recalls, Corrections and Removals (Devices)", FDA, Apr. 25, 2019. https://www.fda. gov/medical-devices/postmarket-requirements-devices/recalls-corrections-and-removals- devices.

[21]   "Medical Device Recalls," *FDA*, Sep. 26, 2018, https://www.fda.gov/medical-devices/ medical-device-safety/medical-device-recalls.

[22]   Fuhr, T., K. George, and J. Pai, "The Business Case for Medical Device Quality," *McKinsey Center for Government*, Oct. 2013, https://www.mckinsey.com/~/media/McKinsey/ dotcom/client_service/Public%20Sector/Regulatory%20excellence/The_business_case_ for_medical_device_quality.ashx.

[23]   Results for 2019 retrieved from the FDA's "Medical Device Recalls Database," https:// www.accessdata.fda.gov/scripts/cdrh/cfdocs/cfRES/res.cfm?start_search=1&event_id=&p roductdescriptiontxt=&productcode=&IVDProducts=&rootCauseText=&recallstatus=& centerclassificationtypetext=&recallnumber=&postdatefrom=&postdateto=&productsho rtreasontxt=vulnerability&firmlegalnam=&PMA_510K_Num=&pnumber=&knumber= &PAGENUM=100.

7

HDO Lifecycle

Although the HDO's perspective on lifecycle management (and cybersecurity in general) is not a main topic for this book, it is important for the manufacturer to understand how the device will be managed by their customers and device operators and how cybersecurity artifacts will be utilized by HDOs. In the most general sense, and as illustrated in Figure 7.1, the main lifecycle management phases from the perspective of the HDO are:

- Preprocurement;
- Procurement;
- Deployment;
- Operations;
- Decommissioning.

These are leading practices adopted by many HDO to address medical device security within their operations. As MDMs interface with these through several critical deliverables, they need to have a good understanding of the HDO lifecycle (Figure 7.1) so to assure proper alignment of activities and expectations.

The HDO lifecycle receives input in the form of actual devices, supporting documentation and artifacts provided by the manufacturer, and risk mitigations during the deployment and operations phases.

Obviously, there can and will be differences in implementation and execution depending on the type of device and use case, type of organization, and whether the device will be based in the hospital or will be in the patient's home.

Figure 7.1   HDO lifecyle.

Another differentiating factor is who will perform the activities. Some or all may be provided by HDO staff, by the device manufacturer, or by a third-party service provider. This may lead to differences in the actual implementation of the process but, more importantly, may create additional communication and process interfaces between the respective activities and between the varying stakeholders, which requires careful consideration so as not to create gaps in responsibility or disconnects in communication. It must be clearly defined which party owns the cybersecurity responsibility and which parts thereof. For example, clinical engineering (or an outsourcing partner) may be responsible for the security of the device itself, but IT may be responsible for maintaining the network. This scenario would require close coordination of not only security responsibilities, but also integration management, network architecture, data flows, the application of security controls, implementation of risk mitigation, and incident response.

This type of a generic device lifecycle has been in use for a long time and its applicability is not limited to networked medical devices. But adding cybersecurity considerations to it requires profound changes to each of the phases, as well as to the roles of the individual contributors.

7.1   Preprocurement Phase

Preprocurement can be broken down in several main tasks by itself. It is a critical phase in cybersecurity planning, as this is the time to:

- Define security expectations as a buying criterion;
- Establish agreement among internal stakeholders.

Preprocurement is, in a sense, the phase where expectations are aligned and where directions are set. It is an opportunity to align priorities among a broad range of stakeholders, including clinicians, clinical engineering, IT and IT security, purchasing, legal and compliance, and to an extent even executive management.

Under the umbrella of preprocurement, we typically find the following activities. Some are more of a general nature and do not apply to a specific device or type, but rather define the criteria for the general, security-aware procurement activities. Others, however, may be specific to a device, device type, or specific purchasing project.

- *Governance:* Based on the overall organization's mission and objectives, governance establishes a framework of cybersecurity rules and practices, inclusive of applicable laws and regulations as well as business criteria, by which leadership defines goals and objectives, establishes responsibilities (duties, privileges, and roles), ensures accountability, proper supervision and control, as well as information-flow and monitoring of implementation and compliance.

- *Policies and procedures*: Based on the established governance, define policies, processes, and procedures as they pertain to the individual medical device cybersecurity lifecycle phases as well as the key risk management and incident response activities.

- *Risk criteria*: Define criteria for cybersecurity risk acceptability as they will be applied as part of the risk assessment.

- *Replacement planning*: Needs and requirements analysis in conjunction with assessment of current inventory, providing guidance on replacement priorities in alignment with clinical and patient care needs. As an overarching process, replacement planning provides the opportunity to align cybersecurity expectations among stakeholders and to balance cybersecurity expectations with clinical needs and budget realities. Replacement planning is also an important long-term planning tool to allow an HDO to align cybersecurity priorities with organizational goals and budget strategy. This process should result in visibility of future security needs as well as engagement of stakeholders in the planning process.

- *Experience analysis*: As an ongoing activity it is important to continually analyze relevant cybersecurity measures, including reported issues with individual devices and device types, performance and compliance of vendors, monitoring of the cyber threat landscape, and analysis of information provided through public resources (e.g., information sharing and analysis organizations; security research).

- *Security requirements definition*: Every HDO should have a clear definition on their security requirements that can be shared with internal and external stakeholders. Although these requirements will establish the objectives required to improve the overall security posture of devices and the device ecosystem, it also has to be rooted in reality and be based on what is practically and economically feasible. Simply establishing a criterion of "a vendor shall deliver devices that are free of vulnerabilities" is neither practical nor possible; therefore, such a requirement is not useful. Better examples have been provided by organizations like Mayo Clinics [1], Indiana University Health [2], the U.S. Department of Defense (DoD) [3], and the Unified Medical Device Cybersecurity Alliance (Kaiser, Cleveland Clinic, Mayo, Froedtert) [4].

- *Training and staffing needs*: Another overarching objective is to assure that medical device cybersecurity activities are sufficiently supported by appropriately trained staff. This includes training users (clinical staff) in cyber-appropriate behavior as well as enabling device maintenance staff to securely deploy and manage devices. The training can be general (e.g., security best practices) or device-specific.

- *Publication and communication*: It is beneficial to all participants, internal and external, if the general cybersecurity expectations are communicated and everybody's expectations can be aligned. The previously provided references are good examples of how health care organizations can not only establish security criteria, but also efficiently communicate them to their vendor community.

- *Requests for vendor input*: An HDO may decide to query their vendor community for input on their degree of cybersecurity compliance (e.g., by requesting vendor disclosure of their security practices and processes). This helps with vendor pre-selection as well as sets the right expectations with the vendor partner.

- *Vendor and partner preselection*: Specific activities may be undertaken to support vendor and partner preselection to build a trusted partner list or in support of a specific procurement project. This could, for example, happen in form of a request for information (RFI) or similar formal or informal procurement tool. This activity can be considered analogous to the MDM's management of an approved supplier list (see Chapter 3).

7.2 Procurement Phase

Procurement activities apply to a specific device or device type in the context of an upgrade, replacement, or expansion project. It is in this phase that an HDO has the opportunity to clearly define security expectations, assess the most suitable vendor and product, and formally seal an agreement, inclusive of cybersecurity requirements, in form of a contract. Typically, procurement includes the following activities:

- *Requirements definition*: An HDO's announcement of its intent to purchase (e.g., in form for a request for proposal (RFP)) should include security requirements, including technical, documentation, testing, communications, and procedural requirements.

- *Device selection*: Based on vendor-provided documentation and artifacts (discussed in detail in Chapter 8), market research, and past experience, the device selection process compares the available offering against the established cybersecurity criteria. The final selection will typically be a balance between budgets, clinical needs, and cybersecurity properties. It should be the intent of an HDO to procure the most secure device possible, however, this should not lead to compromise of the device's clinical features. Typically, the final device should provide security properties that meet the established risk acceptance criteria. Any exception from that should be managed via deviation process and mutual agreement among stakeholders. The device selection process may include a preselection process as well as a final selection step that may include review of provided security documentation, confirmation of those documents and security properties against established risk criteria, and potentially independent validation and testing.

- *Contracting*: As the final step in the procurement process, contracting provides the formal agreement on the device'ssecurity capabilities as well as, on the procedural side, the vendor's maintenance and communication processes. Ideally, the contract should cover both the device security requirements and compliance with FDA Premarket Guidance, as well as the vendor's ongoing support obligations in line with the FDA Postmarket Guidance. The delivery of documentation and artifacts, as described in Chapter 8, should enable the HDO to manage the device through its entire lifecycle: from deployment, to operations (including maintenance and management), through to decommission and end of life (EOL).

7.3   Deployment Phase

Once the new device has been received, the following typical activities take place:

- *Predeployment review and testing*: to assure conformance with requirements and completeness of delivery. Specific tasks may include:
  - *Device software update*: if a new software version has become available since the device was manufactured, it should be upgraded to the latest available level;
  - *Security documentation review*: assure completeness and compliance of provided supporting documentation;
  - *Security testing*: Vulnerability scan, security evaluation, or penetration testing as well as deployment on a test network to assess the presence of potential device security compromise (there have been reported cases of devices delivered with malware already on them);
  - *Remote scanning capabilities*: test device against the organization's commonly used tools like vulnerability scanner, asset discovery, computerized maintenance management system (CMMS), or configuration management database (CMDB) and establish allow/deny rules for the respective tools.
- *Installation*: Deployment of device in its target environment, including security-specific considerations (e.g., implementation of physical security controls).
- *Integration*: In addition to integration with the medical device network and clinical IT systems, the device needs to be integrated with and tested against security-specific tools like: asset management (CMMS, CMDB), vulnerability scanner, cybersecurity management systems (SIEM, SOC), secure remote access, credential management system, and automated risk management systems.
- *Implementation*: The final phase of implementation includes security-specific training for users (as needed) and maintenance staff and ultimately the clinical go-live, at which point the device enters the operations phase.

7.4   Operations Phase

During the operations phase certain security-relevant tasks are performed:

- *Vulnerability and threat monitoring*: The device operator should monitor commonly available sources for information pertaining to newly discovered vulnerabilities or threats. Such information can typically be obtained from the device manufacturer, government agencies (e.g., FDA, DHS, FBI), industry peers, ISAOs, or security researchers.

- *Event and incident monitoring*: From a cybersecurity perspective, one of the most critical tasks is that of ongoing detection and monitoring of security events (an observable occurrence) and incidents (the compromise of an asset with potentially adverse consequences). Such cybersecurity signals may be provided by features of the device itself, through the review of log data, or the observation of the device's environment (e.g., at the network level).

- *Incident response*: Should a new cyber threat or ongoing incident be discovered impacting the device, a specific, predetermined incident response process should be initiated. This process should include decision triage including clinical aspects (e.g., movement of patients), roles and responsibilities, defined technical response steps (identification, containment, eradication, and recovery), internal and external communication, preservation of artifacts, and forensic information. Due to the criticality of our medical device ecosystem, the development, training, and testing of an organization's incident response plan is an often-overlooked but critical task. This is discussed in much more detail elsewhere [5].

- *Ongoing management*: Typically, the device manufacturer provides a preventive maintenance plan, which should include tasks related to maintaining the device's cybersecurity posture. This may include regular review of log files and password changes, as well as checks for the availability of software updates.

- *Deployment of security mitigations*: At times the operator of the device will be made aware of the necessity of security mitigations. In many cases this information will be provided by the manufacturer, but could also come from other sources, including observations by the operator themselves. The required mitigation may be a configuration change, software update (patch), or an external measure like network configuration. In either case, the change must be (1) tested before live deployment in a clinical environment; (2) deployed in alignment with ongoing clinical operations and care delivery priorities (requiring formal change management); (3) evaluated for its impact on dependencies across other systems.

7.5 Decommissioning Phase

HDO cybersecurity activities pertaining to the device's useful clinical EOL should not be confused with cybersecurity activities triggered by EOL considerations on the MDM side. A device's clinical life usually ends when:

- It can no longer safely fulfill its intended function and provide the desired result;
- If it can no longer be maintained;
- It is superseded by a next-generation or better device.

Especially in the latter case, it is common that such devices are resold, donated, or returned to the manufacturer, any of which requires that any sensitive data be purged from the device. This includes any patient information (as defined by local regulators), but should also include hospital- or user-identifying information, as well as device and user credentials. Typical data types that should be removed from any device leaving an organization are:

- Protected health information (PHI) and personally identifiable information (PII);
- Network configuration data including network credentials, host names, IP addresses, firewall configurations, and Wi-Fi passwords;
- User accounts and passwords;
- Clinical trial and research data;
- Location history data (for mobile devices).

The device manufacturer should provide information on

1. The type, amount, and time span of data held by the device;
2. A protocol defining how the critical data should be removed or destroyed in accordance with local regulatory requirements.

In absence of such information, the device owner and operator should take measures to destroy the device or the device's data carriers.

Additional considerations for the specific scenario of device ownership change include possible requirements around license transfer to the new owner. For this purpose, many organizations establish a formal transfer of custody process. The device operator should have formal processes in place to manage device EOL as well as perform the necessary certification and documentation steps.

To summarize, although the main responsibility for proactively securing the device itself lies with the MDM, the HDO has a unique opportunity to improve the security posture of its device ecosystem through the adoption of leading lifecycle management practices. This will allow an HDO to reduce their risks by a) applying mature security processes and technologies, and b) developing a constructive security relationship with their vendors. Failing to do so will force an HDO into a reactive cycle of security fire drills that will neither improve their long-term security posture, nor be resource-effective. A reactive organization is bound to fail its patients.

7.6 Special Scenarios

In the context of HDO cybersecurity lifecycle management, there are two scenarios that require special consideration.

Leased/loaned/demo devices: In principle, the lifecycle steps for any temporary device are the same as described above. However, there are two special considerations:

- The lifecycle phases happen with a much higher frequency than with traditional owned devices, which may create some unique challenges;
- The device owner may not be the same as the operator, and therefore, special considerations with regards to security responsibility and data sanitization may need to be applied.

Legacy devices: Typically, the term legacy applies to devices that as a whole or in part (e.g., a key software component like the operating system) are no longer maintained by the manufacturer. Although this problem also exists outside of cybersecurity (e.g., availability of spare parts), it takes on unique challenges in security as this means that key maintenance strategies, like software patching, are no longer available. This means that complementary security mitigation measures need to be applied externally to the device, using strategies such as network segmentation, firewalls, or anomaly detection systems. It should be noted that this approach may, to an extent, protect against network-based security threats, but will not prevent a security compromise through nonnetwork vectors, such as a data carrier.

7.7 Summary

In this section we discussed the main lifecycle management phases from the HDO perspective: preprocurement, procurement, deployment, operations, and decommissioning. The medical device manufacturer typically interfaces with the procurement phase, which ends with the delivery of the device and its supporting cybersecurity artifacts, and the deployment and operations phases through the delivery of risk mitigation measures. It is important for the MDM to understand how their lifecycle process and phases, interface with the HDO's lifecycle phases and functions.

References

[1]   The Mayo Clinic, "Medical and Research Device Risk Assessment Vendor Packet Instructions," October 23, 2019, https://www.mayoclinic.org/documents/medical-device-vendor-instructions/doc-20389647.

[2]   Indiana University Health, "Information Security - Medical Device Security and Responsible Vulnerability Disclosure," https://cdn.iuhealth.org/resources/Medical-Device-Security-and-Responsible-Vulnerability-Disclosure.pdf.

[3]   Hardman, A., and W. Martin, "Risk Management Framework for DoD Medical Devices," *HIMSS 18*, Las Vegas, NV, Session 136, March 7, 2018, https://365.himss.org/sites/himss365/files/365/handouts/550231590/handout-136.pdf.

[4]   Wirth, A., and C. Falkner, "Cyberinsights: Cybersecurity as a Team Sport," *Biomedical Instrumentation & Technology*, Jan./Feb. 2020, Vol. 54, No. 1, pp. 64–67, https://doi.org/10.2345/0899-8205-54.1.64.

[5]   Wirth, A., and S. Grimes, "Medical Device Cybersecurity: A Guide for HTM Professionals," Association for the Advancement of Medical Instrumentation (AAMI), May 2018, https://my.aami.org/store/SearchResults.aspx?searchterm=mdc&searchoption=ALL

8

Documentation and Artifacts

During the development lifecycle of a medical device, a large amount of documentation needs to be created, reviewed, approved, controlled, and managed. This is not only a good engineering practice, but it is also required by most regulators, for example as found in the U.S. 21 CFR Part 820 Subpart D [1] and ISO 13485:2016 [2]. While there are excellent electronic tools available to perform requirements management (e.g., Jama Connect), it is still fairly common to see small MDMs using standard office productivity suites (e.g., Microsoft Office) as their tools of choice to manage requirements. But no matter if an MDM uses a paper-based approach or a commercial tool, the content represented in this chapter is equally valid for both the simple and more advanced requirements management systems.

Over the course of the last 10 years, we have seen those MDMs who were early adopters of secure medical device development completely change how security artifacts are managed in their organizations. These artifacts are the documentation and output of all the security activities, including plans, reports, test results, and summaries. In the past these artifacts were extremely tightly controlled, with very limited distribution, encrypted and gated with eyes only types of access. During the past decade, however, we have learned that little is gained by such restrictions, while much is lost in the productivity and growth of the organization's teams, the organization's security posture, as well as the customers' opinion of the MDM.

Today, it is common practice to share security artifacts throughout the MDM's organization and even with their third-party contractors. And due to changes in regulatory guidance, much of the content found in these artifacts are now available to regulatory agencies, as well as HDOs and other end users.

We have increasingly realized that keeping security measures secret generally results in worse security implementations, not better. By the same token, sound security measures do not become less secure if stakeholders are aware of them: a locked door remains locked, even if the door manufacturer publishes a statement declaring this door is locked. We are now seeing this principle codified.

8.1 Overview of Secure Development Deliverables

While the following content elements need to be incorporated into the device's design history file (DHF), the vehicle that contains each of these elements certainly does not need to adhere to the structure below. However, this structure does work and aligns with most development activities and flows.

In a sense, these artifacts are an output of the development lifecycle and its supporting processes. When defining and implementing structured processes as described in other chapters, the MDM must anticipate the need for processes to produce artifacts that facilitate auditability, documentation, and transparency.

The following list of artifacts are all created during the development lifecycle (see Chapter 4) and are utilized for development purposes, as well as regulatory submission and future modifications to the device. While all of these elements must be entered into the DHF, as with any other set of DHF artifacts not all will necessarily be required for all activities (e.g., regulatory submission). Submission expectations are being updated in many markets worldwide as of this writing, including the FDA's premarket submission guidelines [3]. Artifacts can include:

- System security plan;
- Design vulnerability assessment;
- System security architecture;
- Interface control document (ICD);
- Testing reports;
- SBOM;
- System security report;
- Labeling, including technical service manuals and instructions for use (IFU).

It is assumed that each of these deliverables contains the usual boilerplate sections, such as purpose, scope, glossary, references, approval, and a traceability matrix where applicable. It should also be mentioned that a risk assessment needs to be performed for the device as a whole and subsequently any newly discovered vulnerabilities that could result in patient harm. While such vulnerabilities would be discovered during the secure development lifecycle, the assessment, triage, supporting documentation, and testing that may result from such discovery are performed by the risk management organization, not by engineers or security experts. Good references for security process and risk management include TIR 57 and TIR 97 from AAMI, and the Joint Security Plan from HSCC.

8.2 System Security Plan

While this has traditionally been a separate document, it is becoming more common for this content to be incorporated into the overall project plan. This document or section defines all the security-related activities required for development, addressing the following attributes for each activity. Such plan may be defined via a corporate standard which covers some or all of these items and would be an indicator of organizational maturity:

- *Description*: What is performed during this activity and what part(s) of the unit under test are being subjected to this activity?
- *Deliverables*: What outputs are created as a result of performing each activity?
  Deliverables typically encompass the following:
  - Design vulnerability assessment;
  - System security architecture;
  - Testing (fuzzing, penetration, static analysis, unit/boundary);
  - SBOM;
  - System security report;
  - Labeling.
- *Responsibilities*: Who is responsible for defining, performing, reviewing, documenting, and approving each activity?
- *Estimated schedule/milestones*: When in the project will each activity be performed?
- *Exceptions*: Are there any anticipated exceptions (such as fuzzing a debug port (JTAG), etc.)?
- *Security goals*: Defines the voice of security high-level abstract goals, similar to voice of customer or voice of business.
  - Nonfunctional goals. Examples include: The medical device system should be developed in accordance with the secure development lifecycle SOP or secure coding conventions should be defined and followed for each of the computer languages utilized in the system;

- Functional goals. Examples include: Maintain confidentiality of data, maintain the integrity of data, or all security mitigating controls shall be FIPS-140 compliant.
- *Cybersecurity risk determination and justification*: The FDA will expect a statement of the product's cybersecurity risk along with a justification for the stated level of risk. Rubrics have been proposed for this, however, at the time of this writing, the exact method that the FDA will be expecting is in flux and should be defined in the final version of the Premarket Cybersecurity Guidance. It will almost certainly not be the standard medical device risk classification (I, II, III) or IEC 62304 software safety classification (A, B, C). Neither one of these standards adequately addresses the uniqueness of cybersecurity risk. As just one example, a medical device could be installed into a hospital's intranet, introducing a vulnerability that allows an attacker to utilize this device as a pivot point into the rest of the hospital's network, without affecting the delivered therapy. This would put other patients at risk and possibly other hospital business processes as well, but may not pose any direct risk to users of the vulnerability-bearing device. Clearly, this type of assessment can be difficult to accurately perform.

8.3   Design Vulnerability Assessment

The design vulnerability assessment includes adaptation of threat modeling tools for system modeling [4], listing of vulnerabilities, screening for false positives, and scoring the remainder. Further details on these activities and their inputs and outputs is provided in Chapter 4.

8.4   System Security Architecture

Developed during the early design phase, this artifact is the result of several other activities:

- Formal decomposition of the system and scoring of the vulnerabilities discovered, as described in Section 8.3 and in detail in Chapter 4;
- Investigating mitigations to be applied to the highest-scoring vulnerabilities.

The artifacts created during decomposition and scoring may be kept for the DHF, but are not typically needed for submission as the results of these activities will be summarized in this architecture document.

This is one of the most important artifacts, as it guides the design and implementation of security into the medical device. Performing this activity early will minimize impacts on the device design and development process, as well as minimizing end-user burden for security controls.

The architecture document should contain:

- Overview of the system topology used for decomposition.
- Cybersecurity hygiene best practices [5].
- A description of the decomposition process, as for example, a description of STRIDE.
- A description of the scoring rubric, as for example, CVSS v2.
- A section containing all the high-scoring vulnerabilities. Each vulnerability should itemize:
  - The vulnerability ID number from the threat modeling tool;
  - The environment of the vulnerability (the area in the system diagram where the vulnerability exists);
  - Name of the vulnerability;
  - Category, in other words, which letter of STRIDE;
  - Description of the vulnerability;
  - Indication of potential impacts to safety and efficacy;
  - The predesign vulnerability score;
  - Trace to mitigation(s) which address this vulnerability.
- A section containing all the recommended mitigations. These mitigations should be highly detailed technical descriptions, including diagrams, algorithms, and libraries to utilize, and should not be limited to software-based mitigations alone. Including:
  - Hardware-based mitigations, such as the selection of a microcontroller that supports hardware acceleration of cryptographic operations;
  - Guidelines for hardening operating systems and other commercial software, as applicable;
  - Physical security controls including side-channel protections;
  - PKI and certificate constraints to avoid a weaker possible selection (e.g., MD5 vs. SH256).

8.5 Interface Control Document

While technically not a security document, an ICD is a very valuable technical document that covers all aspects of a given communications interface, including the technical aspects of security and security mitigations for a given communications medium. Sequence diagram, algorithms, and timings are all frequently utilized in these artifacts.

Tracing is commonly performed from this document to the system security architecture and detailed design documents. This can be an extremely important artifact when addressing questions from the FDA following a premarket submission.

Security-relevant sections in the ICD include:

- Overview;
- Description of the communications medium and standard;
- Assumptions/constraints/risks pertaining to:
  - Environment of operation;
  - Unsupported environments;
  - Expected environmental interferences;
  - Expected performance;
  - Communication quality of service (QoS);
  - Inherent risks of this communication medium.
- Use cases;
- General interface requirements;
  - Security features:
    - Authentication;
    - Authorization;
    - Encryption;
    - Integrity;
    - Antireplay.
- Detailed interface requirements:
  - All interactions detailed to the lowest level of abstraction.

8.6 Testing Reports

8.6.1 Fuzz Testing Report

Fuzz testing (using automation to deliberately feed invalid, unexpected, and random data into the system or software interfaces in attempt to provoke a crash or other unanticipated, possibly exploitable, behavior, and log the results) should be performed as early as possible during the implementation phase of a new medical device development effort. The fuzz testing report should include:

- A description of what interface is being fuzzed and to what level the fuzzing will be performed.

- Target device under test (DUT) identification and version number.
- Name of the fuzz tester.
- The date fuzzing began.
- Description of any specific setup needs, such as previous packets, settings, configuration, or environment.
- Iteration count of the total fuzz testing execution.
- Packet contents of failed fuzzing events.
- Description of the crash or interruption, including any reported assertions, or processor exceptions, and the possible results of such an event. However, at no time should assertions about the impacts on safety and efficacy be made by the fuzz tester or report writer.
- If applicable, citation of relevant CWE ID numbers [6].

8.6.2 Static Analysis Report

Static analysis is a method of computer program review and debugging that is performed by examining the code without executing the program. The process provides an understanding of the code structure and helps to ensure that the code adheres to industry standards. By running the source code through a set of predetermined criteria, static analysis tools expose various types of flaws in the code. Static analysis tools function independently of compilers, linkers, and hardware, and can analyze the most commonly utilized languages, C and C++.

Commercial static analysis tools do not yet replace traditional (human) code review entirely, but they make the code review process more efficient and productive. The next generation of static analysis tools do show promise of improved automation of the code review process, and with further development, complete review automation could become possible in the future.

Coding rules and guidelines exist to ensure that software is:

- *Safe*: usable without causing harm. This is especially critical in medical devices where code supporting functions that would typically be innocuous could produce a ripple effect impacting patient safety.
- *Secure*: its vulnerabilities are mitigated. This is primary focus of SAST (see below).
- *Reliable*: functions as it should, every time.
- *Testable*: can be evaluated.
- *Maintainable*: as the codebase grows, we want to ensure it is designed for maintainability. Coding rules and code analysis are established for many purposes including assuring software quality, but it is important to assure that security objectives are included and met.

We saw an example of unmaintainable code with the Toyota acceleration issue. When Toyota analyzed their dysfunctional code, they found that it had become unmanageably complex over years of continuous development with insufficient checks and oversight [7].

- *Portable*: functionality is the same across various environments and systems.

Coding standards are a collection of coding rules, guidelines, and best practices to improve the safety, quality, and security of the implementation. These may also include guidelines on coding style. Using established coding standards:

- Ensures compliance with ISO, which is required for most major medical device markets;

- Guarantees consistent code quality—no matter who writes the code;

- Secures the software right from project start;

- Reduces costs by speeding up time-to-market through reduced variability and uncertainty, as well as by decreasing overhead for postmarket support.

Static application security testing (SAST) is a subset of static analysis where the emphasis is on security-related vulnerabilities. By applying the appropriate security-related standards and rule set, weak or substandard code with potentially exploitable security vulnerabilities can be identified, triaged, and addressed. The most common standard and rule set utilized is the CERT C and C++ by SEI 2016, which presents hundreds of examples that illustrate coding practices widely acknowledged to represent potential cybersecurity concerns side-by-side with compliant solutions [8].

Ideally, SAST should be performed as often as possible. Given the current state of continuous integration (CI) tools, it is possible to integrate the rulesets and static analysis tools into the build process and have an intermediate and informal result set reported to the developers on an established cadence (nightly builds, weekly builds, etc.).

SAST is expected by worldwide regulators. It is also a best practice to ensure that the source code is properly secured, both from a technical and business aspect. We see many examples of security exploits due to improperly tested code that exact a heavy cost on industry and end-users.

DHF records from SAST activities include the following:

- *Ruleset*: To conduct proper SAST, the development and operations teams should select the appropriate standards and guidelines (i.e., CERT C and then down select the necessary rulesets). Note that it is possible and valid to mute specific checks within the rulesets as long as there is a rationale and justification provided in the ruleset record, which should be preserved in the DHF.

- *Initial assessment results*: Using the rulesets, an initial run of SAST is used to Identify all the violations. The premitigation result should be recorded in the DHF. Note that it is possible and valid to tweak previously selected and previously muted checks as long as there is a rationale and justification provided. Also remember to update the ruleset records (above) appropriately.
- *Postmitigation results*: The premitigation results are used to triage the identified violations. These fall into three categories:
  - True positives that must be addressed; these require changes to the code, refactoring, or even possible design.
  - True positives that can be suppressed; these typically apply to low-risk violations that do not provide the necessary risk/benefit. These are referred to as residual risk elements.
  - False positives that are either suppressed or muted; violations identified but do not apply, in which case they can be suppressed or the check itself can be muted (this is only if the violation is globally false).

    After all the violations have been triaged and addressed, the post-mitigation results are also recorded in the DHF. Note that the triage and address of violation should not be a one-type exercise and should be executed and repeated as often as practical throughout the development life cycle process, with records captured for each instance.
- *Final SAST report*: The SAST artifact needed for submission is the final SAST report. This report collects and summarizes all three preceding activities. It provides the rationale and justification for the standards and rulesets selected and muted, the pre-mitigation results, and postmitigation results, including any suppressions and residual risk elements.

8.6.3 Penetration Testing Report

Penetration testing is a (friendly) hacker-for-hire service where a security expert attacks the medical device under development. Penetration testing can be performed at two different times in the development lifecycle: concurrent with the implementation phase, and/or at the end of all development activities. If performed concurrently with implementation, then penetration testing can consist of several iterations, each scoped to specific areas of the medical device. This is sometimes referred to as µPen or micropenetration testing. This is the preferred approach to penetration testing because results are delivered while the implementation phase is still in progress require fewer resources to address compared to one-time, post-development penetration testing, which delivers results after all development activities have concluded.

The penetrating testing report should include:

- A description of the penetration testing effort.

- The scope of the penetration testing (full system or specific subsystems in an μPen testing scenario).

- Target device under test identification and version number.

- Name of the penetration tester.

- Date of the penetration testing.

- Equipment utilized in the penetration testing.

- Description of any specific setup needs such as communications, settings, configuration, or environment.

- Description of the results of the penetration effort, including the possible implications of such an attack. However, at no time should assertions about the impacts on safety and efficacy be made by the penetration tester or report writer.

- If applicable, citation of relevant CWE ID numbers [6].

8.6.4   Boundary Testing Reports

Unit testing is conducted on the smallest testable components of the software. In procedural languages like C, this is an individual function in isolation. In object-oriented programming, this may be an entire class or interface. Boundary testing is a subset of one coverage metric of unit testing; this specific metric challenges the units on parameter boundaries to ensure the proper decision and branching is executed and the parameter boundaries are not violated within the unit. Like SAST, unit boundary testing should be performed as often as practical; ideally, that means written and executed in parallel with actual development. While this is not practical in all development life cycles, at minimum each unit should have an associated test case by the time prototype implementation is complete.

Boundary testing is also required by regulators of most significant markets worldwide. It should also be a best practice to ensure that each unit is tested appropriately for its functional behaviors. This testing is critical for functions that have parameterized inputs that require a specific range or even specific discrete values. The functions must ensure that they can handle and manage out-of-bound values properly through decision and/or branch logic. Another critical aspect is to ensure that modified conditions (i.e., parameters with valid in-bound values that are changed or updated with the function) are taken into account.

The following artifacts should be provided as inputs to unit boundary testing:

- *Test case*: Each test case exercises a specific functional or behavioral path within a unit (function) for a given module. Test cases should be completely independent of each other. If you have to combine test cases or suites in any manner, they are not being executed properly.

- *Test suite*: Collection of orthogonal tests cases for a given module.

- *Mocks and stubs*: Mocks or stubs are specially generated functions that replace the actual function calls from the unit under test because they reside outside the module in scope. This gives the test developer flexibility to add/modify the stub functionality to inject the necessary stress for a given test scenario or behavior. Mocking and stubbing also allows removal of hardware dependencies and requirements for testing.

- *Test harness*: Collection of test suites, stubs, and mocks along with test validation functionality. This is tool-dependent.

- *Test runtime*: Tool-specific run time executable or library. This is also tool-dependent.

Finally, the unit boundary test report is the output artifact. It summarizes all test results, postmitigation only (initial results with premitigation failures are generally not required). This is because the understanding and expectation of boundary testing is that all failures are addressed with robust test coverage and fixes to the implementation (code) itself. The reporting structure itself is tool-dependent and proprietary, as well as custom-fitted to each project implementation.

8.7 SBOMs

Knowing what we are getting when we buy something is a foundational concept of the modern marketplace. Every piece of candy one buys from the supermarket, for example, comes with a list of ingredients. Most of the time, we may not care about the specifics of what is enrobed in creamy chocolate. We may not care, that is, until someone in our life has a serious food sensitivity. In this case, it is imperative that the data about these ingredients is available, in an accessible format. Why is it that we expect more transparency from a twinkie than from the software on which our medical devices depend?

Building on the model of a bill of materials in manufacturing, an SBOM amounts to a list of ingredients that make up software. Just as every manufacturing process begins by determining the necessary underlying parts, a good software development process should explicitly track what components are used, too. Producing an SBOM is a reflection of the MDM's security maturity and sharing an SBOM extends that security commitment to the customer to enable them to understand the composition of the product. This also gives the end-user insight into the potential risks of discovered vulnerabilities that may be disclosed for items in the SBOM.

Software is seldom written from scratch in today's digital economy. Instead, we use existing libraries and modules to take advantage of the efficient functionality that others have developed, either in the form of open source software libraries, or licensed commercial software from others.

This, in turn, introduces risk. Unlike custom software, third party components are more widely deployed, therefore, are often more familiar to malicious actors and are more targeted. When vulnerabilities are discovered in a third-party component, an SBOM will help all involved understand whether or not a specific device is at risk.

8.7.1   Elements of an SBOM

An SBOM is a nested inventory of third-party software components as well as the MDM's application software. A recent international initiative is underway to better define SBOMs as identifying and listing software components, information about those components, and the supply chain relationships between them [9]. The SBOM enables visibility and clarity into the complexity of the underlying components that make up modern software. Information about these components can be mapped to other sources of data, which in turn supports better decision making.

The basic unit of an SBOM is the software component. A component means a third-party software component (TPSC), or any identifiable subdependencies that make up that TPSC, such as a software library, that originate or could be used outside the primary organization. Components may contain more subcomponents that, in turn, may be made of more components. This level of specificity allows us to get more granular than, say, a package. On the other hand, it does not require the complexity and public detail of, say, a particular function.

Components should be sufficiently uniquely identifiable, as should the SBOM at its highest level. There are several objectives for identifying components. First, two SBOMs referring to the same component should be able to be certain that they are truly referring to the same component. As such, a cryptographic hash should be used to uniquely identify components. Second, the component identified must be able to be mapped to other sources of data about the software component, such as to a vulnerability database, or a set of data about software licenses.

These components are organized in a dependency tree, documenting sub-components that are included in a component. Alice's application may depend on Bob's browser and Bingo's buffer. Bob's browser, in turn, may include Carol's compression engine. Ideally, this set of dependencies would include all sub-components, until primitives are identified that have no further dependencies. In reality, obtaining this level of transparency may not always be possible. The necessary depth will be a function of market demand, and of the cost of obtaining that information.

8.7.2 SBOM Formats

As a relatively new security concept in many industries, including health care, there is no single broadly accepted mechanism for defining or handling SBOMs. Several data formats exist that can be used for defining TPSC data in machine-readable formats, including SWID, SPDX, and CycloneDX. Each of these can capture the basics of component information. The format selected should follow the relative advantages of each standard [10].

An SBOM entry should be created to describe each component. In cases when the producer does not have complete information, an SBOM can explicitly indicate the distinction between a known full set of underlying components versus incomplete information. When the product changes, or when an underlying component changes, the SBOM should always change as well. In cases when a software supplier may discover something new about the underlying components, the SBOM should reflect that new information.

Given the range of implementations around SBOMs across the diversity of the software ecosystem, there is no single way to share SBOMs. In general, data should flow downstream from the supplier to the user. In the modern, complex supply chain, there may be many supplier/user dyads, as each user in turn becomes a supplier to others. Still, the channels for sharing can be reasonably scoped. In some cases, the data can simply follow the source code or binaries. In other instances, such as on a medical device with constrained resources and limited exposure to scanning tools, the data should be shared through some other mechanism. The manufacturer usage descriptor (MUD), a secure set of on-device URLs, has been suggested as an excellent mechanism [11].

8.7.3 SBOM Applications and Use

While recent regulation may call for SBOM generation in particular instances, there is real value in tracking and sharing third party dependencies across the supply chain. Some use cases are clear, such as the ability to detect vulnerabilities in underlying components discussed above. Others are more nuanced, such as having insights into when a key dependency may be nearing its end-of-life or is otherwise no longer supported as expected.

One particularly valuable approach to understanding the benefits of transparency in the software process is to better understand the roles that we may play in the supply chain [12]. Those who make products that include software should, one would hope, have some understanding of the code that they are shipping in their products. Knowing about potential vulnerabilities is not only good for security, but can increase efficiency: when a new vulnerability emerges, a development team lacking an SBOM would have to review all their software to determine if any of it has a problem. Operationally, it can support the use of blocklists or allowlists of banned or approved components, respectively. Lastly, many organizations already evaluate some aspects of the code they use in order to comply with different types of intellectual property licenses.

SBOM data can also help those who select software for their organizations, and those who run the software. In selecting software, whether purchasing it on devices, or choosing to run an open source package, SBOMs can help identify potentially vulnerable components or highlight areas where more targeted security analysis is needed. Once selected, the software must be operated. Devices with SBOM data are easier and more effective to operate. When a new vulnerability is discovered, the operator can more easily make a determination about whether that device is affected. If there is some risk, SBOM data can give an earlier warning, enabling independent mitigations while waiting for a patch—assuming that a patch is, indeed, forthcoming.

The impact of greater transparency extends beyond individual organizations. When more actors in the marketplace are aware of what is being used, what is being avoided, and why, it can drive market pressure to use better components. When developers and purchasers pay attention to components, those selecting them are more likely to select them with greater care, asking important questions about the quality and security of the ingredients they use to assemble their products.

Despite the rather evident benefits of transparency, SBOMs are still, in 2020, an emergent deliverable. Much of the initial work has emphasized the benefits of a light-weight approach. This supports scalability for easier and faster adoption and flexibility across a range of organizations and development environments. The time-honored approach of rough consensus and running code will allow early adopters to innovate, and others to learn lessons from their experience. A lightweight approach supports further extensions, as we learn to bundle SBOM data with other security and supply chain metadata we care about without breaking the end-to-end approach across the complex supply chain.

8.7.4   SBOM Artifacts

SBOMs support multiple use cases, including:

1. Exposing to an HDO the software components being utilized in a medical device for rapid automatic identification of medical devices potentially at risk due to a disclosed vulnerability. Note that this still leaves the HDO with the challenge of understanding whether a disclosed vulnerability is actually exploitable in the given implementation, an assessment for which they will typically need the support of the MDM.

2. Exposing to an HDO the software components being utilized in a medical device prior to sales of the device to an HDO as part of the incentive for an HDO to purchase the medical device.

3. Managing the security posture of software components being utilized in an MDM's engineering organization for the life of a product.

4. Managing the licensing posture of software components being used in an organization's products.

5. Complying with regulatory expectations in supplying a written SBOM as part of premarket submission.

6. Exposing to end-users the software components being utilized in a medical device (see the section on labeling).

Given the large number of software components being utilized in modern medical device systems, items 1 through 4 require the use of machine-readable SBOMs, while items 5 and 6 depend on written SBOMs. The difference in medium impacts SBOM content. To perform items 1 through 4 well, the SBOM needs to contain all of the components being used, including each component's dependencies. As an example, the current version of Log4j (a popular logging library) has 294 subdependencies. These can be included in a machine-readable SBOM, but in a written version of the same SBOM, a single layer of component depth is sufficient and anything beyond that would not be practical—including secondary levels would overwhelm the reader with too much information. In this example, Log4j would be listed in the written SBOM, but its subdependencies would need to be included in the machine-readable SBOM.

In machine-readable SBOMs, the subdependency layers of included software components should be traced to the maximum discoverable depth, thus exposing dependencies and vulnerabilities that may otherwise be overlooked. For a good example of this read about Urgent/11—IPnet was so far down in the dependency tree that it was difficult for manufacturers to determine whether their product was at risk [13]!

8.7.5 Remaining SBOM Challenges

When fully realized in the marketplace, SBOMs will be a very valuable tool for securing medical devices and their environment. However, there remain some challenges that will need to be overcome before this can happen. These include:

1. Naming the supplier;
2. Naming the component;
3. Identifying the device's version (down to a security-relevant level);
4. Creation of a machine-readable SBOM;
5. A mechanism for the MDM to expose the degree of exploitability for each TPSC relative to its specific scope and use in their device;
6. A mechanism for the MDM to automatically ingest SBOM information and map newly disclosed vulnerabilities via the SBOM to the individual device.

Of these, the hardest are the first two. Unique naming is a difficult solution without cost pressures or a central authority that can be queried to correlate ACME and ACME Inc. Likewise, there is the need to support aliases and indications to mark a supplier that has been acquired by another organization with a different name. No one solution will be adequate for this task, so a flexible approach will have to be agreed upon that can be easily and quickly processed by an automated asset management system for medical devices. Fortunately, all of these issues are being investigated by the NTIA's SBOM software component transparency working groups [14].

8.8 System Security Report

This report is created at the end of all development and cybersecurity related activities. The purpose of this report is to summarize for senior management and regulatory professionals the outcome of all security activities performed during the development of the medical device. The system security report contains:

- A reiteration of the cybersecurity risk determination and justification found in the security plan artifact.
- Similar to the security plan, this report details each of the performed activities with summarized results including:
  - The number of vulnerabilities discovered;
  - The number of vulnerabilities mitigated;
  - The number of residual vulnerabilities, including a description and a risk-benefit analysis justifying the residual status for each one.
- The SBOM: an itemized list of an application and its TPSC being utilized in the medical device, cross-referenced to a recognized security dis-

closure database such as the National Vulnerability Database (NVD) [15]. The system security report must include justification for any TPSC on the SBOM with known vulnerabilities in terms of each TPSC's specific use within the device, the exploitability of those vulnerabilities, and the severity of the risk to end-users and the intended environment (HDO, home, etc.).

MDM senior management must review and approve this artifact before it can be considered complete. This demonstrates senior management's awareness of the risks and potential attack surface of the new medical device.

8.9 Labeling

According to the FDA's October 2018 premarket guidance draft [3], from which the items below are derived, device labeling must include the following (labeling includes package, part, and document labels; technical manuals; instructions for use; promotional material including websites; and statements made by the MDM) [16]:

- Device instructions and product specifications related to recommended cybersecurity controls appropriate for the intended use environment (e.g., antivirus software, use of a firewall).

- A description of the device features that protect critical functionality, even when the device's cybersecurity has been compromised.

- A description of backup and restore features and procedures to regain configurations.

- Specific guidance to users regarding secure operations and supporting infrastructure requirements so that the device can operate as intended.

- A description of how the device is or can be hardened using secure configuration. Secure configurations may include endpoint protections such as anti-malware, firewall/firewall rules, allowlisting, security event parameters, logging parameters, and physical security detection.

- A list of network ports and other interfaces that are expected to receive and/or send data, and a description of port functionality and whether the ports are incoming or outgoing (note that unused ports should be disabled).

- A description of systematic procedures for authorized users to download version-identifiable software and firmware from the manufacturer.

- A description of how the design enables the device to announce when anomalous conditions are detected (i.e., security events and incidents).

Types of security events that the device should detect and announce include configuration changes, network anomalies, login attempts, anomalous traffic (e.g., send requests to unknown entities).

- A description of how forensic evidence is captured, including but not limited to any log files kept for a security event. Log files descriptions should include how and where the log file is located, stored, recycled, archived, and how it could be consumed by automated analysis software (e.g., IDS).
- A description of the methods for retention and recovery of device configuration by an authenticated privileged user.
- Sufficiently detailed system diagrams for end-users.
- An SBOM, including but not limited to a list of commercial, open source, or off-the-shelf applications, libraries, frameworks, and operating systems to enable device users (including patients, providers, and HDOs to effectively manage their assets, to understand the potential impact of identified vulnerabilities to the device and any connected systems, and to deploy countermeasures to maintain the device's essential performance.
- Where appropriate, technical instructions to permit secure network (connected) deployment and servicing, and instructions for users on how to respond upon detection of a cybersecurity vulnerability or incident.
- Information, if known, concerning device cybersecurity end of support. At the end of support, a manufacturer may no longer be able to reasonably provide security patches or software updates. If the device remains in service following the end of support, the cybersecurity risks for end-users can be expected to increase over time.

References

[1] "Title 21: Food and Drugs, PART 820—Quality System Regulation, Subpart D—Document Controls," *U.S. Code of Federal Regulations*, Apr. 13, 2020, https://www.ecfr.gov/cgi-bin/text-idx?SID=df2431a9085035a24294a4709b05253c&mc=true&node=sp21.8.820.d&rgn=div6.

[2] International Standards Organization, "ISO 13485:2016 Medical devices—Quality management systems—Requirements for Regulatory Purposes," Mar. 2016, https://www.iso.org/standard/59752.html.

[3] "Content of Premarket Submissions for Management of Cybersecurity in Medical Devices; Draft Guidance for Industry and Food and Drug Administration Staff," *FDA Guidance Documents*, Oct. 18, 2018, https://

www.fda.gov/regulatory-information/search-fda-guidance-documents/content-premarket-submissions-management-cybersecurity-medical-devices.

[4] Shostack, A., *Threat Modeling: Designing for Security*, Wiley, Hoboken: NJ., Feb. 17, 2014.

[5] https://cybersecurityforum.com/cybersecurity-faq/what-is-cyber-hygiene.html.

[6] https://cwe.mitre.org/.

[7] Safety Research & Strategies, Inc., "Toyota Unintended Acceleration and the Big Bowl of "Spaghetti" Code," Nov. 7, 2013, https://www.safetyresearch.net/blog/articles/toyota-unintended-acceleration-and-big-bowl-%E2%80%9Cspaghetti%E2%80%9D-code.

[8] SEI CERT, "SEI CERT C Coding Standard," 2016, https://resources.sei.cmu.edu/downloads/secure-coding/assets/sei-cert-c-coding-standard-2016-v01.pdf.

[9] National Telecommunications and Information Administration Framing Working Group, "Framing Software Component Transparency: Establishing a Common Software Bill of Material (SBOM)," Nov. 12, 2019, https://www.ntia.gov/files/ntia/publications/framingsbom_20191112.pdf.

[10] National Telecommunications and Information Administration Open Working Group, "Survey of Existing SBOM Formats and Standards," Oct. 25, 2019, https://www.ntia.gov/files/ntia/publications/ntia_sbom_formats_and_standards_whitepaper_-_version_20191025.pdf.

[11] Lear, E.,R. Droms, and D. Romascanu, "Manufacturer Usage Description Specification," *Internet Engineering Task Force*, Mar. 2019, https://tools.ietf.org/html/rfc8520. *See also* https://www.mudmaker.org/.

[12] National Telecommunications and Information Administration Open Working Group, "Roles and Benefits for SBOM Across the Supply Chain," Nov. 8, 2019, https://www.ntia.gov/files/ntia/publications/ntia_sbom_use_cases_roles_benefits-nov2019.pdf.

[13] U.S. Food & Drug Administration, "URGENT/11 Cybersecurity Vulnerabilities in a Widely-Used Third-Party Software Component May Introduce Risks During Use of Certain Medical Devices," *FDA Safety Communications*, Oct. 1, 2019, https://www.fda.gov/medical-devices/safety-communications/urgent11-cybersecurity-vulnerabilities-widely-used-third-party-software-component-may-introduce.

[14] https://www.ntia.doc.gov/SoftwareTransparency.

[15] https://nvd.nist.gov/.

[16] https://www.fda.gov/medical-devices/overview-device-regulation/device-labeling.

9

Organizational Development of Roles and Responsibilities

Obviously, there is no single approach to defining the right organizational model for a cyber-capable manufacturer, nor is there a set of well-defined roles and responsibilities that would work for every manufacturer, every device type, and every target market. However, we can make some general assumptions about certain functions that need to be provided that then will map to respective roles and responsibilities. The particular organization's need will determine if all of these roles are required, whether they should be dedicated or if they can be fulfilled part time, if they can be extensions of existing roles or should be created as new and dedicates positions, or if they can even be contracted to a third party.

This section will provide some insight and suggestions for building a device manufacturer's cybersecurity organization. As such we will focus on product security, with limited discussion of traditional IT and IT security roles as applicable to product security and internal infrastructure and communication needs. For example, a manufacturer's support organization may take on specific security-related tasks in their relationship with their customers, and for that they may have certain manufacturer-internal IT needs like VPN access, strong authentication, and similar.

In addition to touching on the role of the individual and his/her roles and responsibilities, we will also discuss the role and benefits of working with external stakeholders (like information-sharing organizations) that play key parts in solving the medical device cybersecurity challenge.

9.1   Roles and Responsibilities: Overview and Rationale

Recognizing the complexity of today's medical devices and medical device eco-system, as well as the unique cybersecurity challenges facing the industry, it is clear that well-defined roles filled by skilled and experienced staff are of utmost importance. Furthermore, the ability to effectively and constructively communicate between internal and external stakeholders, within the scope of the defined roles, is a prerequisite for success.

The understanding and proper execution of cybersecurity responsibilities goes up the chain of command, starting with the individual engineer, to engineering management, regulatory and legal, and all the way up to executive leadership. This complex distribution of roles requires not only the clear identification of responsibilities, but also continual education on medical device cybersecurity, as we are facing a constantly evolving field with new cyber threats and new regulatory requirements emerging continually.

Building a cybersecurity-capable organization can be quite a challenge, though. Based on conversations we have had with manufacturers that did so, we would assume that in most cases, 50% to 75% of the new security positions will have to be filled with new hires in order to obtain the right skill set and shoulder the additional workload. To add to the challenge, the required skill sets are currently not easy to be found in software engineers and other required roles, although we would hope that this will improve in the future.

It is essential to establish clearly defined roles and responsibilities, ensuring that stakeholders can meet what is required of them as well as allowing for the integration of individual functions within the larger ecosystem. A common approach to accomplish and communicate this is the so-called RACI matrix, which helps identify and define activities for which stakeholders are responsible, accountable, consulted, and informed, as shown as an example in Table 9.1.

The responsibilities of the manufacturer-internal roles span the range from enabling the organization to execute (typically senior management responsibilities like budgets, staffing, enablement), to defining the right security requirements (driven by regulations and shifting market needs) and implementation (software design best practices and applicable security technologies), to providing a proactively secured product to the market, together with the appropriate regulatory approval and documentation artifacts that are useful to the customer.

Furthermore, large or midsized manufacturers may benefit from separating the roles into two categories: cybersecurity responsibilities of dedicated roles as part of a central organization and cybersecurity responsibilities (dedicated or not) as part of existing engineering or similar roles. In a sense, this would be a horizontal overlay with general, mainly strategic responsibilities that is differentiated from vertical and more tactical roles that focus on a specific product or

Table 9.1
Example of a High-Level RACI Matrix

Responsibility	Role											
	Executive Leadership	Product Cybersecurity Executive	Regulatory and Legal	IT and IT Security	Sales and Sales Operations	Product Management	Engineering Management	Architecture / Systems Eng.	Security Engineering	Software Quality	Sustaining (Security) Eng.	Service (maintenance & support)
Establish security governance and budgets	R/A	C	C	I	I	I	I	I	I	I	I	I
Identify regulatory objectives	I	C	R/A	I	C	C	C	I	I	I	I	I
Assure staffing and training levels	I	R/A	I	I	C	C	C	I	I	I	I	I
Security Requirements	-	C	C	I	I	R/A	C	C	C	C	I	I
Risk Management	I	C	C	I	C	C	C	C	R/A	I	I	I
Develop security architecture	I	I	I	I	I	C	C	R/A	C	C	I	I
Secure code design	-	-	I	-	-	I	I	C	R/A	C	I	I
Code review	-	-	-	-	-	I	I	C	C	R/A	I	I
Verification & Validation	-	I	I	-	-	I	I	C	C	R/A	I	I
Manufacturing Transfer	-	I	I	-	-	I	I	C	C	I	R/A	I
Customer Communication	C	R/A	C	-	C	C	C	C	C	I	I	I
Security Maintenance	-	I	I	-	I	R/A	C	C	C	I	I	C
Incident Response	I	C	C	-	C	C	C	C	C	I	I	R/A

R: Responsible
A: Accountable
C: Consulting
I: Informed

part of the process. The advantage would be the separation of responsibilities and avoidance of conflicting objectives, whereas the disadvantage could be challenges in alignment of goals and communication.

In that sense, and recognizing that each organization is unique and will likely find its own approach on how to organize around cybersecurity, here is some general guidance on the respective responsibilities:

- *Executive leadership*: It all starts at the top. Although executive leadership typically does not own responsibility for specific products, it does own responsibility for the company's general priorities and direction. It needs to have strong influencing skills, the ear of company management, and some degree of authority so to constructively influence direction. It is executive leadership's role to establish a culture of security and enable the business case for security. This should also include defining governance, providing strategic direction, defining risk tolerance and business cases, establishing the right organizational structure, and funding the execution through budgets and staffing. Leadership's involvement in security matters should be continuous in order to understand, and correct as needed, gaps and respond to developments that may require attention (e.g., security incidents reported by customers).

- *Product cybersecurity executive role*: Some, especially larger, manufacturers may find it beneficial to create a dedicated strategic security role, for example a Chief Product Security Officer, that will be responsible for the overall program and implementation as well as the management of the company's security strategy. Especially with large organizations that need to align their security objectives and approach across multiple product lines, this type of role is essential.

  Furthermore, such a senior executive will also provide an important external-facing function by engaging with industry, regulators, and customers. In other words, he/she is the public face of the company on all cybersecurity matters as it relates to the company's products as well as public representation on cybersecurity topics.

- *Regulatory and legal*: Cybersecurity is a developing topic that, over the past few years, has been added to the responsibilities of regulatory/quality/legal roles. Understanding the evolving regulatory requirements and providing guidance to the organization (leadership and engineering) is critical as we are still far from a settled regulatory landscape and final (stable) requirements.

  In essence, new cybersecurity tasks for regulatory and legal departments include: understanding developing international regulations, inclusion of cybersecurity processes in the software quality management system, establishing cyber-relevant complaint management and regulatory reporting, the adoption of applicable legal requirements and their inclusion in the company's contract framework, and review of customer contracts with regard to security issues in those contracts.

Further tasks may be required in addition, such as: support of regulatory filings and approvals, response to customer inquiries, review of contracts, and supporting the preparation (and if needed delivery) of communication to customers, regulators, and the public in response to a cyber incident.

Larger companies may benefit from a dedicated cyber security standards specialist—a person or team focused on and knowledgeable in the evolving cybersecurity standards and responsible for translating these standards into policies and procedures as well as defining and managing internal practices for implementation across various functions including engineering and services.

- *IT and IT security*: Manufacturer-internal IT/IT security's involvement in product security issues is typically peripheral. However, there are some manufacturers that have established direct or dotted line reporting of product security into IT security, presumably to leverage expertise. Also, there are some areas where IT / IT security involvement typically is required:

  - Assure security and integrity of the manufacturing environment so that neither the manufactured devices nor the manufacturing equipment (IT gear, ICS, robots, etc.) are at risk of a security compromise. Any security incident impacting the manufacturing environment can result in compromised devices or device software or compromise of manufacturing quality, either of which may result in costly remediation and recalls. For example, according to a survey conducted in late 2018 by the Health Information Management and Systems Society (HIMSS), 7% of hospitals have reported receiving hardware or software infected with malware off the shelf (e.g., preloaded malicious software) [1].

  - Assure security and integrity of the service (maintenance and support) IT infrastructure, including: providing tools for secure remote access to customer-based devices through tools like VPN or multifactor authentication; assuring security of field personnel IT equipment (laptops, data carriers); and protecting manufacturer IT infrastructure (local and hosted) that holds customer data.

  - Provide and maintain a secure environment for engineering software tools and data (e.g., documentation, source code) to assure confidentiality (e.g., protection of intellectual property), integrity (e.g., protect source code repositories), and availability (e.g., backup and recovery).

- As needed, support IT integration with as well as review and audit of third-party partners and suppliers.

Note that if service activities involve access to PHI or if PHI is transferred to the manufacturer's IT infrastructure as part of a service provided, under the U.S. Health Insurance Portability and Accountability Act (HIPAA), the manufacturer will most likely be considered a Business Associate (BA, defined as any third-party that receives, transmits, maintains, or creates PHI on behalf of a HIPAA covered entity). This would then require contracting (BA Agreement), HIPAA-compliant IT infrastructure for transmission and storage of PHI, and HIPAA training of employees that interact with or have access to PHI. To determine whether an MDM would need to comply with HIPAA, we advise seeking appropriate legal counsel [2].

- *Sales and sales operations*: As the face to the customer, sales staff needs to be educated on cybersecurity in general, the company's overall security strategy and value statements, as well as the security particulars of the individual products. Furthermore, although sales may not be responsible for providing security content and answer security-related technical questions, they are certainly the communication channel for providing security information to the customer and for responding to their inquiries, be it through a formal process like RFI/P, contract, or MDS^2, or through an informal process of general customer questions.

What may complicate this particular area of organizational and cultural change is the complexity of many sales organizations that may have to serve different countries (and need to deliver different, regional security messages), or in other cases where the sales organization may be one level removed when a manufacturer uses a reseller in certain markets.

Larger organizations may create a dedicated back office role or team to respond to customer inquiries, fill out security questionnaires, provide security documentation, and similar functions.

- *Product management*: As the typical owner of a product, product management's main responsibilities with regards to cybersecurity are to analyze and understand the evolving international regulatory requirements as well as changing customer needs. These can be aligned with the company's business objectives (e.g., in which global markets to sell the product), resulting in the development and management of product security requirements specifications and roadmaps as input to the engineering process. Product management also plays a vital role in supporting regulatory filings, internal education, customer communication, and postmarket management.

- *Engineering management:* The main role of engineering management is one of enablement to assure that the MDM's engineering organization can successfully execute on delivering secure devices as per defined requirements. This is a combination of budget management, resource planning and staffing, providing the right tools and complementary services, and enabling education on the topic. Furthermore, engineering management should assure compliance with corporate governance and product specification and plays a key role in internal stakeholder communication and key process steps like manufacturing transfer. Lastly, it is typically engineering management's role to triage events and help develop an action plan to address the engineering part of the problem—be it a customer-reported incident or an internal issue during product development.

  One important role of engineering management is the maintenance of cybersecurity information, be it the source code itself or the documentation artifacts created for regulatory filing, internal documentation, or to share with customers. Although the documentation requirements, formats, and methods may be defined by others (e.g., by a quality organization, developing and managing the systems that collect product security information and support process execution) is of critical importance and part of the engineering organization's responsibility. Artifacts that need to be managed, some for internal and some for external purposes, can include risk analysis, threat modeling, security controls, MDS^2 and supporting data, SBOM, as well as information on vulnerabilities and mitigation (e.g., updates and patches).

- *Security systems engineering:* The art of systems engineering consists of achieving an optimum balance between all elements in the system. This can be especially difficult when each element has different attributes with different competing priorities; compounding this, changes to any single attribute may change another element's attributes, causing a ripple effect throughout the design, prompting systematic reassessment. This delicate refactoring and releveling can be performed over and over again in the effort to seek the correct balance in the system, where "correct" itself may be a changing goal. Today's regulatory and customer expectations means that the balance of priorities familiar to traditional systems engineering must also include cybersecurity considerations.

A hypothetical example of competing priorities in a development project is an implantable medical device that needs to physically fit in a sealed titanium enclosure with extremely limited physical battery space; yet the implantable device needs to have a battery life measured in years, due to surgery being required to gain physical access to the battery (with all the attendant risks accompanying that surgery). To achieve this long battery life the CPU needs to be in a low power mode for as much time as possible, occasionally coming up to full power to perform its principal activity. If you have a hardware-accelerated cryptographic engine, utilizing cryptography only adds microseconds to the full power operation; yet trying to perform these same operations purely in software would consume hundreds of milliseconds of additional full-power operation, obviously taking a significant toll on the battery life.

The placement of security mitigations in the design must be purposely performed to address specific weaknesses and risks, and the electronic hardware needs to be included in this so that features such as hardware-accelerated cryptography will be present in the design. Otherwise, the design may require a bigger battery and bigger case; or else the designer must accept a shorter battery life leading to more frequent surgical replacement.

Once this balance has been achieved, it is the responsibility of the security systems engineer to formalize this balanced system through the use of interface control documents (ICD). This type of document describes the relationship that holds all the elements of the system together, including communications, functionality, timing, and security mitigations. It is a highly technical document containing sequence diagrams, data flows, details of packet contents, and so on. The ICD is the ultimate authority on the activities of the communications. It should be written at a level that an average staff software engineer can implement the specified design.

When defining the topology of the system, the final, complete system should be stipulated, even if the device's rollout is in a phased deliverable approach where the initial version of the system may not contain all elements that will be supported in future releases or in various configurations of the system. A good example of this is to support firmware updating via a communications medium only, even where there are plans to roll out additional functionality over that communications medium in future versions.

The system-of-systems elements are not necessarily defined by physical boundaries, but by logical grouping at a given level of abstraction. Furthermore, in that systems-of-systems, the individual components may have different levels of regulatory controls. An example would be that commonly with a picture archiving and communication system (PACS) the diagnostic viewing software is regulated, but the actual workstation, backend servers, and storage are not regulated.

Security Architecture/Systems Engineering

Whether security architecture is a separate role or part of systems engineering is typically dependent on the company (e.g., size), system architecture (e.g., complexity), and use case (e.g., patient risk). In either case, a separate and dedicated security architecture effort early on in the design phase is critical to assure that the right security decisions are made and that security objectives are met.

This includes applying security engineering best practices to the architecture itself (device internal as well as target integration architecture), support of the selection of complementary security technologies and tools, definition of the right security controls and metrics, and monitoring design deviations relative to the requirements during the development phase. Furthermore, security architecture and systems engineering typically play an important role in regulatory filings and the creation of documentation and artifacts.

Security Engineering

Again, several factors determine whether security engineering is a separate role or a function being provided by security-capable software engineers (if available). Either way, this function is critical to assure a secure final product that is supported by secure development activities that follow established software design best practices and standards. This includes the development itself as well as support of testing, regulatory filing, documentation development, manufacturing transfer, and, on an as-needed basis, postmarket issues. Also, typically as a task conducted in cooperation with security architecture, this role supports critical security analytics tasks as for example security risk assessment or threat modeling.

Software Quality

In the context of cybersecurity, quality assurance's responsibility is similar to traditional QA. This includes testing (discussed in detail in Chapter 8), verification and validation activities, and producing supporting documentation prior to and as part of manufacturing transfer. However, what is unique for the security processes is that these reviews and testing happen much earlier in the engineering process and is, in many cases, a continual or iterative process, rather than a distinct activity that happens at the end of the development phase. Another difference is that during the course of the secure development lifecycle, cybersecurity issues that potentially affect patient safety and efficacy will be referred into the software QA process for risk-based evaluation and determination of potential impacts.

Since cybersecurity is a unique problem area requiring a unique skill set and the use of specialized tools, organizations may benefit from a dedicated cybersecurity tester (or group) that is knowledgeable on various testing techniques including automated and manual, along with interpretation and disposition of results. Typical tools applied to assure secure and quality software are code review, fuzz testing, pen testing, vulnerability scanning, and similar (discussed in more detail in Chapter 8).

Out of similar considerations, the function of vulnerability management may require a dedicated resource that has the right skill set and is dedicated to managing the intake, tracking, assessing, and facilitating response to newly identified vulnerabilities

Sustaining (Security) Engineering

This is a common continuing support function that typically takes on maintenance engineering responsibilities after manufacturing transfer. This includes support of the initial transfer, the resolutions of any issues that may occur during this process, as well as management and synchronization of future updates and issues that may manifest during manufacturing.

Service (Maintenance and Support)

Every manufacturer has some form of maintenance/support/service function. In the context of cybersecurity this can be broken down into two main sets of responsibilities:

- Support customers with any security issues that may arise (e.g., incident response) and help with the management of security issues like deployment of updates, communication of issues, and so on.
- Assure that any service or maintenance activities do not pose a security risk to the customer environment. This includes, but is not limited to, secure remote access (VPN, multifactor authentication, etc.) as well as preventing on-site transfer of malware from service equipment (laptop, USB, etc.) into the customer environment.

Other, general responsibilities for service and maintenance are the escalation of any customer security issues as well as compliance with HIPAA (or other local regulations) requirements in handling customer equipment or data.

The above is a somewhat generic view of the commonly required roles. As stated before, every manufacturer, device, use case, or market have different requirements and require a different approach and capabilities. These roles may be developed as a new and dedicated function but also could be an overlay to or matrixed into already existing roles, or a mix of the two approaches. Also, every manufacturer (and individual employee) starts at a different level of security capability and a careful plan should be developed to form a sensible path forward. Although cybersecurity is critically important, moving towards a cyber-capable organization cannot be rushed. Understanding the current state and development of a step-by-step plan to reach the desired final state are essential.

Also, there are certain tasks that would span across several of these roles and that are discussed in more detail elsewhere in this book. For example, supplier and supply chain risk assessment and management, response to customer questionnaires and contracts, vulnerability disclosure, or incident response. Because these can occur and impact any (or multiple) of the phases in the development lifecycle, a cross-functional approach is typically required.

9.2 Training and Education

Each of the roles reviewed above require appropriate security training and education. In today's medical device world, nobody should be exempt from keeping their cybersecurity knowledge up-to-date. Obviously, the need for continual education would be very different across the roles discussed above, be they technical or nontechnical, or engineering or administrative. For example, an executive decision maker or regulatory staff would need quite a different perspective than an engineer who writes code.

In cybersecurity, education is a continuous process as security technology and best practices, security research, the threat landscape, and regulatory requirements continuously evolve. Security education should be a well-planned and sufficiently funded aspect of any organization. This can happen through participation in conferences, providing access to literature and online resources, or educational courses and certifications provided by security industry organizations or software platform providers.

The following is a list of organizations, several of which provide conferences and events, that the authors found useful to support general and medical device specific cybersecurity education needs:

- H-ISAC: webinars, working groups, and annual spring and fall workshops;
- HIMSS: online resources, webinars, as well as HIMSS Annual Conference;
- AAMI: online resources, webinars, as well as annual AAMI Exchange;
- Health Technology Alliance (HTA, a cooperation of HIMSS, AAMI, and ACCE);
- AdvaMed: online resources, webinars, MedTech Conferences;
- Archimedes 101 Conference;
- Q1 Productions: various events focused on select topics (regulatory, security engineering);
- MedTech Intelligence: events and webinars.

Software platform vendor provided resources:

- Almost every provider of commercial software solutions has published hardening guidelines that should be followed. Especially critical system components like platform (operating system), databases, or software tools for access, authorization, and authentication should never be ignored.

Organizations that provide security education and certifications:

- SANS Institute: providing training courses and certifications on various security subspecialties (e.g., cyber and network defenses, penetration testing, incident response, digital forensics, auditing);
- ISC²: provider of security training and certification, including Healthcare Information Security and Privacy Practitioner (HCISPP) and Certified Information Systems Security Professional (CISSP) certifications;
- Center for Internet Security (CIS): publisher of Critical Security Controls (CSC) cybersecurity tools, best practices, and education, including hardening guidelines, benchmarks, and metrics;
- Information Systems Audit and Control Association (ISACA): a professional association focusing on IT governance as well as providing some well-recognized certifications like Certified Information Systems Auditor (CISA), Certified Information Security Manager (CISM), Certified in Risk and Information Systems Control (CRISC);
- CompTIA: a provider of a range of foundational and advanced IT and security training and certifications, including Network+ and Security+ certifications for beginners;
- Open Web Application Security Project (OWASP): providing free articles, methodologies, documentation, tools, and technologies in the field of web application security;
- National Association of Corporate Directors (NACD): Board- and director-level resources on cybersecurity oversight and responsibility.

Government:

- National Institute for Standard and Technology (NIST): a wide range of publications on software, IT, and security topics;
- National Cybersecurity Center of Excellence (NCCoE): providing security technology as well as medical device specific security use case examples;
- Department of Homeland Security (DHS): a number of security resources, including general and medical device security advisories (ICS-CERT) and security education (NICCS);

- European Union Agency for Cybersecurity (ENISA): general and topic-specific cybersecurity resources;
- European Telecommunications Standards Institute (ETSI): general and topic-specific cybersecurity resources;

Literature and online resources:

- NACD Director's Handbook on Cyber-Risk Oversight;
- Health Care Industry Cybersecurity Task Force: Report on Improving Cybersecurity in the Health Care Industry;
- Healthcare and Public Health Sector Coordinating Council (HSCC): Health Industry Cybersecurity Practices: Managing Threats and Protecting Patients;
- Healthcare and Public Health Sector Coordinating Council (HSCC): Joint Security Plan (JSP);
- CyBok–The Cyber Security Body of Knowledge (free download), University of Bristol in cooperation with several other leading European and U.S. academic institutions;
- CIS Security Benchmark Mapping Guidance;
- OWASP Secure Medical Device Deployment Standard;
- IEEE Computer Society: Building Code for Medical Device Software Security.

There are many more resources than can be listed here, so it is worthwhile keeping an eye on the developing security educational space. Also, we should not undervalue the benefits of internal learning. Many organizations have at least a few people that were early adopters and have been dealing with cybersecurity for a number of years. These employees have developed valuable skills, especially product- and market-specific knowledge that would be difficult to find in public courses or material. It is strongly advised that organizations leverage these experienced people and make internal teaching/learning part of their cybersecurity journey.

It is a generally accepted best practice to make a cybersecurity skills assessment (as appropriate for the role) part of annual personnel assessments or performance reviews, as well as to include specific cybersecurity goals and activities in staff members' annual development goals and training plan.

9.3 Communication

As should be apparent from the previous section, cybersecurity-related tasks span across many different roles across an organization as well as outside of it.

Therefore, communication that keeps all stakeholders in the loop (as appropriate for their roles) is essential to success. This includes communication on both planned tasks and unplanned events.

Planned tasks range from an organization developing high-level cybersecurity goals and objectives, to the specific project plans and milestones that relate to cybersecurity. Unplanned events include disruptions that may occur through cybersecurity incidents, newly released research or vulnerabilities, changing regulations, or changes in the threat landscape.

9.3.1   Internal

Internal communication requirements are best addressed in alignment with existing practices, be it on corporate strategy or in the form of regular project reviews and hand-off sessions between teams at the appropriate project milestones. However, bear in mind that cybersecurity matters may be more dynamic and may require a more continuous approach than is traditionally in place. Team members need to be kept informed about changes in strategy, policy, regulation, security research, or threats as well as reported customer incidents. Review of the lifecycle rings (see Figure 3.1) illustrate the internal communications required on a continuous basis, such as between postmarket surveillance and those responsible for maintaining the organization's ASL. Open communication channels between engineers, security personnel, and legal may also needed when issues or attacks have been discovered.

9.3.2   Customer and External-Facing

Similarly, customer or other external-facing communication is of critical importance to the success of any cybersecurity program. Security-mature MDMs will communicate proactively, in a timely fashion, and in sufficient detail. Unfortunately, past examples have shown that there is still room for improvement and that the willingness to communicate on cybersecurity topics varies widely between manufacturers. A common reluctance stems from the longstanding misconception that sharing security information may be perceived as an admission of poor product quality or become a regulatory and legal liability, rather than being seen as a sign of maturity and responsible participation in community-wide sharing of risk information, as increasingly encouraged by international regulators.

However, leading manufacturers have demonstrated that a well-executed communication process is a sign of process (and cultural) maturity rather than a weakness. This is further encouraged through the FDA cybersecurity postmarket guidance, which provides incentives in the form of eased regulatory reporting requirements for manufacturers to share and collaborate.

In the past, there have been instances where poor vulnerability management has resulted in significant problems for the MDM when, for example, security researchers ended up going to the press because they could not reach anybody within the manufacturer's organization that was willing to listen. Or, in another example, a manufacturer received flak not for the fact that there was a weakness in their product, but because they badly mishandled the vulnerability disclosure and communication. In the present day it is commonly understood that software will always be imperfect, and openness and mature management of vulnerabilities has become the accepted path to a more mature state. But as with all matters related to medical device cybersecurity, the industry is still learning and improving practices.

We should also not overlook the fact that external communication is not limited to what goes on between the MDM, client HDOs, and the FDA or other regulators. It may, at times, include other government agencies, law enforcement, the press, as well as patients and caregivers [3].

Following are descriptions of three complementary communication models for addressing particular scenarios.

Scenario 1: Vulnerability Sharing

Mature MDMs have established and published what is commonly referred to as responsible vulnerability disclosure policy. This should cover the two most prevalent scenarios:

1. The reporting of a newly discovered vulnerability by a security researcher, customer, or other party to the medical device manufacturer;

2. The communication about that vulnerability to customers, along with a recommended mitigation and advice or procedure for assessment of risk.

The associated processes include definition of handling and processing of vulnerability information, including technicalities like the submission process and channels (phone number, email address, public encryption key, information to be provided). Such a process would provide triage and decision criteria on how a vulnerability is analyzed and how its risk (exploitability, severity) will be assessed. Some manufacturers even have gone so far as to create a Hall of Honors to acknowledge security researchers that have been responsible partners and have helped make their products better.

On the outbound side, the process will define who will be responsible for public communication and how the process will unfold. It will define, depending on the risk level, what action will be required (regulatory filing, customer or public notification, etc.) and how recommendations and priorities will be established.

It is important for any MDM to have such processes defined beforehand (i.e., before a first vulnerability will need to be managed), to have teams with defined responsibilities in place, communications drafted for various plausible scenarios, and processes regularly practiced in mock exercises (with lessons learned promptly implemented). The supporting teams are typically cross-functional and include executives, regulatory, legal, security engineering, and communication resources. Furthermore, it may be required to add external expertise on an as-needed basis—for example a clinical consultant or customer representative to help assess clinical or customer impact.

There are general standards available that can be applied and adopted to the specific medical device situation, for example ISO/IEC 29147:2018 (vulnerability disclosure) or ISO 30111 (vulnerability handling).

Scenario 2: Threat and Incident Response

As a reminder, a threat is merely the potential for a security exploit (e.g., a newly discovered malware), analogous to what we call a hazard in traditional safety risk analysis, whereas an incident is the actual occurrence of a compromise, analogous to harm in safety risk analysis. Neither requires that an actual harmful event has occurred—for example, a security sensor on a hospital network may pick up unusual traffic from a medical device, but the device functionality has not (yet) been compromised. Obviously, if an incident includes harm to patients or users, then this would make it a high-priority item. An MDM may be informed about a new threat that could impact its devices, or a customer may report an actual security incident. Either way, the response process would be similar: (a) support the customer in response and recovery, and (b) analyze and triage the issue so that risk and required action (e.g., disclosure to regulators or the larger customer base) can be assessed. The team that provides such support will most likely be multidisciplinary including executive decision makers, regulatory and legal, engineering, and customer support.

Typical steps included are:

- Customer support;
- Preparation;
- Identification (detection and analysis of incidents);
- Containment and mitigation;
- Eradication;
- Recovery;
- Follow-up (postincident activities and reviews).

By the follow-up step (and often before), the MDM should launch its internal processes (not all steps are required in all cases):

- Triage and initial risk assessment;
- Regulatory and legal assessment;
- Engineering assessment;
- Clinical impact assessment (if needed);
- Initial communication;
- Final risk assessment;
- Determination of mitigation;
- Regulatory and legal action;
- Final communication.

Guidance on how to build an incident response plan and team can be found from:

- ISO/IEC 27035:2016: Information Security Incident Management;
- SANS: Creating and Managing an Incident Response Team;
- CMSEI: Handbook for Computer Security Incident Response Teams (CSIRTs);
- NIST 800-61: Computer Security Incident Handling Guide;
- ISACA: Incident Management and Response;
- NIST Cybersecurity Framework.

Scenario 3: Security Information Sharing

For the purpose of efficient and controlled vulnerability, threat, and incident sharing, manufacturers should utilize ISAOs. Initially established by the U.S. Department of Homeland Security under Executive Order 13691, ISAOs provides a central resource for gathering information on cyber threats to critical infrastructure and facilitates two-way sharing of information between the private and public sector.

ISAOs exist across all critical infrastructure industries, including health care and public health. Although the initial concept around ISAOs was developed in the United States, they (or similar models) have been established in other regions as well. Their mandate includes member education, networking, working groups, as well as vulnerability, threat, and incident information sharing.

Specifically, for medical devices, the FDA has established incentives in their 2016 cybersecurity postmarket guidance for manufacturers to participate in an ISAO. As stated in the document, "the Agency considers voluntary participation in an ISAO a critical component of a medical device manufacturer's comprehensive proactive approach to management of postmarket cybersecurity threats and vulnerabilities and a significant step towards assuring the ongoing safety and effectiveness of marketed medical devices. For companies that actively participate in such a program, and follow other recommendations in this guidance, the Agency does not intend to enforce certain reporting requirements" [4].

Specific examples of ISAOs that have been formed to support medical device cyber information sharing are the Health Information Sharing and Analysis Center (H-ISAC), Sensato, and MedISAO.

References

[1]   HIMSS, "2019 HIMSS Cybersecurity Survey," Chicago, IL, 2019, p. 6, https://www.himss.org/2019-himss-cybersecurity-survey.

[2]   Rose, R., "How Does HIPAA and the HITECH Act Impact Medical Device and Pharma Companies?" *Becker's Hospital Review*, Jan. 11, 2013 https://www.beckershospitalreview.com/healthcare-information-technology/how-does-hipaa-and-the-hitech-act-impact-medical-device-and-pharma-companies.html.

[3]   Abernathy, A., "Balancing Patient Engagement and Awareness with Medical Device Cybersecurity," *FDA Voices*, Nov. 14, 2019, https://www.fda.gov/news-events/fda-voices-perspectives-fda-leadership-and-experts/balancing-patient-engagement-and-awareness-medical-device-cybersecurity.

[4]   "Postmarket Management of Cybersecurity in Medical Devices; Guidance for Industry and Food and Drug Administration Staff," *FDA Guidance Documents*, Dec. 28, 2016, p. 8, https://www.fda.gov/regulatory-information/search-fda-guidance-documents/postmarket-management-cybersecurity-medical-devices.

10

Security Technology, Tools, and Practices

Considering the range of medical device types, variations in architecture, available power and computing resources, and use cases, it is difficult to discuss specific security technologies as they may or may not fit specific devices and implementations. Obviously, a highly resource-constrained and battery-powered implantable device has very different security capabilities and needs than a room-filling piece of imaging equipment that is a system of systems by itself.

Furthermore, we often find a potential conflict between ease of use of clinical features on one hand and security on the other. Therefore, it is of critical importance that the correct security technology is chosen and implemented as appropriate for the device and use case to facilitate usability in a secure environment without compromising clinical features. In some cases, security requirements may need to be subordinate relative to clinical requirements (e.g., a break glass emergency access feature).

Security technology, as available through commercial security products, can be used as a complement to secure design best practices. They should not, however, be used to replace security best practices but rather to provide complementary protection or to address specific problem areas, for example, to protect the inherent vulnerabilities that may come with a commercial operating system.

This section will review security technologies in general terms and discuss the respective technology's advantages and disadvantages in the context of some general implementation scenarios. The terminology describing various security technologies is by no means standardized and different vendors may describe their products and features with different terms. Table 10.1, outlining the main

Table 10.1

Overview of Common Security Technologies Found in the IT Environment

Compliance and Infrastructure Management	
• Compliance & vulnerability managment• Risk management• Configuration managment database (CMDB)	
Endpoint	**Network**
Endpoint protection:	Intrusion detection & prevention (IDS/IPS)
Mobile and IoT security	Network access control (NAC)
Endpoint detection & response (EDR)	Virtual private network (VPN)
Configuration and patch management	Deception (honeypots)
Cryptography	Anomaly detection
Perimeter	**Cloud**
Firewalls	Cloud access and security broker (CASM)
Unified threat management (UTM)	Zero trust platform
IDS/IPS	Cloud-specific security solutions (workload
Web filter/gateway	protection, DLP, authentication, …)
Web isolation	
Enterprise Security	
• Authentication• User behavior analytics (UBA)• Simulation & awareness• Email security • Access control• Encryption & obfuscation• Data loss prevention (DLP)• Web security	
Orchestration and Response	
• Security information & event managment (SIEM)• Security operations center (SOC)• Incident response & remediation	

security technologies in a typical IT environment, is by no means an absolute definition.

The remainder of this chapter will discuss a select number of these security technologies in depth, with focus on endpoint security technology and its application to medical devices.

10.1   Endpoint Security

In traditional IT security, the term endpoint designates either end of a data flow on a network. In that sense, a server is an endpoint (as the origin or depository of data), as is a desktop or laptop computer or mobile device where the user receives or enters data. This statement implies that an endpoint is typically a composition of commercial software components like operating system (OS), databases, web browser, or email program, which then leads to a common feature set typically found in commercial endpoint security products.

Obviously, a medical device is not a desktop computer. But some medical devices are architecturally close enough to a traditional endpoint that these commercial endpoint security products can be applied, either out of the box or with a degree of configuration to tune it for device requirements and restraints.

Commercial endpoint security products usually provide a combination of detection and prevention features as well as offer integration with external security management systems like a SIEM, SOC, or automated EDR.

10.1.1 Antimalware

The best-known and most-used endpoint security technology is traditional antimalware (also called antivirus). Today's antimalware has come a long way from the early days of scan- and signature-based virus detection and incorporates many more features. This is in response to the explosion of malware and malware variants that started in the early 2010's. To use a simple measure as an example, threat research suggests that through the use of automation and tools for virus generation and obfuscation, the number of new malware and variants grew from 1 million a year in 2008 to over 1 million a day since 2014 (but has leveled off in that range since).

Today's commercial antimalware products usually contain a set of features that may include:

- Signature-based scanning (the traditional antivirus);
- Reputation analysis (files, web sites, domains, and so on);
- Behavior analysis (detecting events that are typically associated with malware penetrating a system);
- Critical resource protection (e.g., system registry);
- Network and port controls (e.g., USB autorun);
- Artificial intelligence/machine learning (AI/ML) based features;
- Features addressing specific problem areas through, for example: runtime isolation, exploit prevention, account management, or software-based firewalls.

This list highlights one of the main challenges of using commercial antimalware on medical devices. Although today's solutions are resource-optimized, their impact on performance and behavior of a dedicated purpose system, like a medical device, may be too unpredictable. Some devices, though, may be sufficiently computer-like that this is acceptable.

Some other concerns of using antimalware in the medical device use case are:

- Continual network connection to provide signature updates and to utilize enhanced features like file reputation analysis (which requires comparing file signatures to a reputation database, typically cloud-based).
- Risk of false positives, which, although a rare nuisance on a desktop computer, could have significant impact on a medical device.

- Since antimalware is a reactive technology, new features in response to evolving attack techniques may be more frequent than device updates. Or, such new features may not be compatible with the device and therefore, preventing the device from improving in response to these new attack techniques.

- Moreover, antimalware technology features (and consequently product versions) typically develop faster than medical devices are updated. MDMs may find it challenging to maintain compatibility and assure the latest security features are available for their devices.

- May require operating system integrity and currency, and may not work effectively when using older, less frequently patched, or customized OSs.

Yet, in some cases and with more computer-like devices, antimalware may be a viable choice of security technology as it provides the following advantages:

- It is typically a known technology to the HDO and therefore faces a lower acceptance threshold (or may even be asked-for in bids and contracts).

- HDOs typically standardize on a specific antimalware solution and therefore it may be preferential as it integrates well with their security management environment.

- Easy to install and easy to deploy (but device-specific configuration may be required).

- As these are commercially available products, licensing can be delegated to the HDO.

Whether the antimalware product is MDM-supplied or HDO-supplied, in most cases the MDM would need to verify that the anti-malware product does not adversely affect the device. Based on that, the MDM would then typically supply installation and maintenance instructions for the anti-malware product.

It is a reasonable approach to remove or disable (through configuration) certain antimalware features should they conflict with device performance or features. For example, assuming that most medical devices do not provide (and should not need) email or external web browsing, it is probably perfectly acceptable to turn off these dedicated features in the antimalware product to reduce overhead and improve predictability. Other configuration steps may include restricting resource-heavy tasks like signature update or disk scan to certain times or to less-critical operating states of the device.

However, caution should be taken that the antimalware product is not configured to an extent that its actual security capabilities are compromised.

For example, turning off certain advanced detection or prevention features may reduce the security capabilities of the software (as these advanced features are there for a reason). Or, making the assumption that certain file types or directories are not at risk and excluding them from scanning may be fine today, but will this still be correct after 10 or more years of system operation?

To summarize, reasons that MDMs may select antimalware for their products are: (a) the device is sufficiently computer-like that this is a valid design decision, (b) it is a known entity to HDOs, often in line with their enterprise security standards and procurement requirements, (c) it can delegate license purchase and management to the HDO, and (d) it allows the MDM to check the security box. The latter may be a less desirable (and defendable) rationale and MDMs should be encouraged to look at more appropriate security technologies, as discussed in the next section.

10.1.2 Host Intrusion Detection and Prevention Systems

If we take the (admittedly simplified) view that anti-malware is blocklisting, then host intrusion detection and prevention systems (HIDS/HIPS) can be introduced as allowlisting (i.e., rather than trying to detect certain unwanted behavior like antimalware does, these products control and allow only known good behavior and forbid everything else). This, too, is a simplified description—both types of products have evolved over the years and in today's commercial products have some overlap in security features.

The advantage of antimalware is that it is a perfect fit for the common desktop and server environments where there are certain commonalities in architecture, yet high variance in the individual device as every user is running different programs and performs different tasks. The ability to be easily deployed and managed across such variance is designed into the products. But this comes with significant overhead.

In the medical device environment, we have the opposite situation. Every device of a specific model is identically configured when it leaves the factory and normally there is the desire to keep each device just that way. This is the perfect scenario for using HIDS/HIPS. Although it does require up-front engineering effort for each model, once that is done, it will perform very well on the device.

Typically, implementing a HIDS/HIPS requires configuration of policies that defines allowed system behavior. This includes executable and process permissions, communication behavior, port access, and more.

Once the allowlist policies have been defined, they only require change when the system is updated. Practically, that means that any HIDS/HIPS policy changes can be applied together with the system update (or patch) and no changes are required in between.

One limitation should be mentioned, and that is if a device needs to be configured locally (network configuration or installation of additional software), these changes may be difficult to be anticipated when developing the security policies for the device.

Specifically, the generally recognized advantages of HIDS/HIPS (relative to antimalware) are:

- Lower resource consumption and small storage footprint (typically a fraction), providing predictable behavior and minimal resource impact;

- Tight system control and prevention of changes by malicious actors or unintentional changes due to operator mistakes;

- Does not require signature updates (policy changes can be supplied at the time of software update), therefore supporting disconnected use cases;

- Less dependent on operating system currency;

- As it is not a reactive security technology, it is typically even effective against unknown and zero-day exploits;

- In addition to protecting OS and other commercial products, it can equally protect custom applications and files;

- Enables fine-tuning of system resource allocation as well as provides granular access control;

- Reduced manufacturer lifecycle management pressure (fewer patches);

- Better control of release schedule;

- In many cases, can be applied as a compensating control beyond device end-of-support horizon;

- Improved reliability through system constraints, privilege management, and tamper protection.

In addition to application (or executable, or process) allowlisting and behavior control, commercial products provide: network, port, device, and file controls; software firewall; sandboxing and memory controls; intrusion and exploit detection/prevention features, user and configuration controls; as well as security incident logging and audit support.

Today's commercial HIDS/HIPS products contain features to make implementation easier, for example by providing out-of-the-box policies for typical OS environments or by providing a learn mode to observe normal system behavior (including custom processes) and then turn these into predefined policies. Such a policy baseline takes care of the majority of the configuration work and can be fine-tuned further as part of the engineering process.

From the manufacturer perspective, HIDS/HIPS not only enables more secure devices protected from malicious or nonintentional (well-meaning insider) change, but also reduces lifecycle management pressures (fewer patches are required) and allows the manufacturer to gain better control of the release schedule and commercial aspects associated with updates and patches.

10.2 Cryptography

Cryptography is used to protect various types of systems and their data across many industries and use-cases. Cryptography-based technologies are being used to protect the confidentiality, integrity, and authenticity of information—not just financial or protected health information (PHI), but also less obvious information that flows between devices and applications to perform technical functions in the background. In a sense, information has become the lifeblood of many industries and without it, businesses would come to a halt. However, medical devices have unique requirements when it comes to incorporating cryptography into devices and applications.

Although cryptography is an integral technology that can be used to improve a device's security posture, any cryptography used must carefully consider the impact to the end user along with the device's functionality and reliability. A balance must be achieved between the exploitability and potential impact of a device security incident versus the added complexity, impact on system resources, usability, reliability, failure modes, and functional impact introduced by the cryptographic features.

If a medical device becomes harder to use due to complicated cryptographic authentication mechanisms, that medical device could become unusable at a critical moment. Long and strong passwords are hard to remember and devices used to supply multifactor authentication keys are easy to lose, hard to replace, and can cause healthcare workflows to slow down or even cease completely if it prevents users' access to the medical device.

If a device decides that a critical update is needed in order to continue to function securely, then a patient can be left waiting for the device to become available again at a critical moment in their care. For instance, if a vital-sign monitor used during surgery restarts to perform a security update in the middle of a procedure, vital information would become unavailable to the doctors and nurses when it is needed most.

A pacemaker can use cryptography to verify the settings loaded onto it to ensure that they are intact and have been created by a trusted party. While this seems like a sensible safety mechanism at first, if the settings are rejected due an expired public-key certificate (something that happens quite often with general cryptography systems), then the patient may not be able to get a settings update to their pacemaker to adjust critical therapy functions.

These are just a few examples of how cryptography can have negative, unintended side-effects. However, if implemented with specific medical device use-cases in mind, the end result can be a more secure, and therefore safer, device that is not as vulnerable to hacking, can operate properly in different, potentially hostile, network or physical environments, and can even perform more predictably due to increased trust of the data that drives the medical device's operation.

It is worth noting that there are other forms of security that can be just as important. Cryptography provides information protection and can be complemented by physical security, platform security (e.g., protecting the operating system), and network security, for example. Those forms of security have been addressed elsewhere in this chapter and should be used to complement cryptography.

The focus here will be on three main data security concepts facilitated by cryptography: confidentiality, integrity, and authenticity. These concepts help protect data and applications from being compromised. The data being protected can include anything that the device gathers, generates, stores, or transmits while it is actively performing tasks for a patient, is sitting idle, or is completely offline. The applications being protected can range from small firmware images running on microcontrollers up to full-featured operating systems and applications running on devices that are architecturally similar to general purpose computers.

- The confidentiality of medical device data is achieved through encryption. By encrypting data, we can ensure that it remains private and is only accessible to those who are intended to have access to it.

- The integrity of medical device data can be determined via cryptographic hash functions. By hashing a particular set of data, and storing the resulting hash, the hash can always be recalculated and, if the recalculated hash matches the original hash, then there is high confidence that data has not been modified.

- The authenticity of medical device data can be verified by digital signatures. By signing data with a digital signature algorithm, and sharing the resulting signature, another party can verify this signature to ensure the data came from the original source and has not been tampered with.

Depending on the device type and use case, some or all of the above may be required.

It is not the intent of this section to provide an in-depth review of cryptography engineering, as that has been done elsewhere [1]. It will rather focus on providing a foundation and introduction to specific implementation and maintenance challenges an MDM will encounter due to the unique medical device use case.

There are a number of caveats which concern the effectiveness of encryption, hashing, and digital signing. For instance, just to name a few, cryptography can be broken if the secret key is leaked, [2] if the underlying algorithms have a weakness [3], if a random number generator is not random enough [4], or if there is a side-channel that leaks secret key information [5]. These are only a few of the potential weaknesses and common implementation errors in cryptography. As always in cybersecurity, there is no such thing as perfectly secure, but a risk-based approach should lead to a solution that is acceptably secure for the given use-case.

It is worth reiterating that no system, at least with commercially reasonable effort, can ever be 100% secure. The ultimate hope is a system that is impractical to compromise because the effort would be too high for interested attackers to break in. Even this is a moving target because, as compute power increases (e.g., through the advent of quantum computers), some algorithms in use today may become obsolete because they will be too easy to break. As a result, they may no longer be deemed secure enough for a given application and use case. Consequently, crypto algorithms need to evolve along with the emerging weaknesses [6]. Cybersecurity, unfortunately, is an arms race.

10.2.1 Key Concepts in Cryptography

Properly securing medical devices through the use of cryptography can be challenging. There are many cryptography software libraries and hardware implementations to choose from, each with its own set of features and configuration options with inherent advantages and disadvantages. Also, building an ecosystem around the device that enables the proper management of cryptographic keys (over the lifetime of the device) to facilitate cryptography operations is nontrivial.

In order to secure medical device data with cryptography there are a few common, core cryptographic concepts that are required. At the heart of modern cryptography are asymmetric keys, used for public-key cryptography, and symmetric keys, used to enable encryption. Asymmetric keys enable digital signing algorithms for proving data is intact and was signed by a particular key or party. Symmetric keys enable encryption that is used to encode data in order to keep it private [7].

These cryptographic tools can be used to enable more complex capabilities like a root-of-trust, to prove that certain keys, and their digital signatures, came from a trusted party. In addition to the root-of-trust, special algorithms, like key-exchange algorithms, can be used to securely establish symmetric encryption keys between two parties so that they can send data back and forth privately.

10.2.1.1   Root-of-Trust

A root-of-trust is essentially a hierarchy of asymmetric encryption keys, usually long-term keys (LTKs), whose trust is established through public key infrastructure (PKI) by signing one another's public-key material to form a chain [8]. These signatures, in the form of public-key certificates, provide a high level of confidence that the keys in the hierarchy have vouched for the keys beneath them. This indicates that the keys can be trusted, much like a military chain of command. In a chain of command, each person knows who they trust as their commanding officer and, if their commanding officer issues a command, they are aware that their commanding officer was either given that command by someone higher up in the chain or has the authority to issue that command themselves.

See Figure 10.1 for a simple example of a root-of-trust key hierarchy where each node in the hierarchy vouches for the one below it by signing a digital certificate that creates a trust-chain.

In a root-of-trust, we determine the chain of command, or trust hierarchy, by evaluating what are known as public-key certificates. Among other things, these certificates contain metadata about an asymmetric cryptographic key pair, the public key of the key pair, and the certificate source or issuer. Since we know that a cryptographic signature can only be constructed by the secret key in an asymmetric key pair, if the signature can be verified, then we know that the information in the certificate has been vouched for by the key that signed the certificate. Usually, the signature in a certificate is generated by the key that is one level higher in the chain from the public key that is in the certificate. For instance, the certificate for the medical device's keys in Figure 10.1 would be

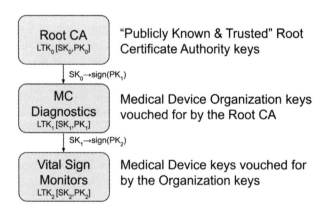

Figure 10.1   Simple root-of-trust key hierarchy.

signed by the organization key, thus establishing that the device's keys should be trusted by the organization. In this example, there is one special certificate that is signed differently from the rest. The first key, or root key, has no key above it in the hierarchy. This root key signs its own certificate, known as a self-signed certificate, and establishes what is called a root certificate. This root certificate establishes the ultimate trusted key in the hierarchy. As long as a root certificate is stored in a secure location on a medical device, where it cannot be tampered with, then this root certificate can be used as a trusted starting point, or trust anchor, to determine the level of trust of any other key that has a certificate in the root-of-trust hierarchy.

Once trust of the asymmetric cryptographic key-pairs can be established, that trust can be extended to data that has been signed or encrypted by those keys. Verifying digital signatures is a fairly straightforward concept. Given some data and a signature for that data, a cryptographic operation can verify whether the signature, which is generated by a particular public-key, matches the data. If the signature can be verified then there's a high confidence that the data was signed, or vouched for, by that particular key. Therefore, if a certain key is trusted, based on its certificate-chain in the root-of-trust, then the data signed by that key can also be trusted.

10.2.1.2   Encryption within a Root-of-Trust

Once we have a root-of-trust and can establish which keys are trusted, as well as what data is trusted through digital signatures, we can extend that trust to data that is encrypted in various ways. At the heart of data encryption is a symmetric key that can be used to both encrypt and decrypt the data. Even though data is encrypted, the original data can still be vulnerable to a leak if the symmetric encryption key is not properly protected. The trick to keeping encrypted data as secure as possible is choosing a proper strategy for establishing the symmetric encryption key, protecting it, and using an encryption algorithm that does not have weaknesses that are exploitable by potential attackers.

There are a few common strategies for establishing symmetric keys between two parties so that they can encrypt and decrypt data for one another. Some strategies are more straightforward but have serious weaknesses that can compromise the security, and some strategies require more involved procedures to establish the symmetric keys but have desirable security benefits.

10.2.1.3   Preshared Keys

Preshared symmetric encryption keys are the easiest to understand. A symmetric key can be created, much like generating a password, and that key can then be given to any parties that wish to encrypt and decrypt data for one another. While this is easy to understand, it is also easy to identify weaknesses. If the distribution of these keys between the parties is insecure, then the keys can be easily compromised from the start. If this key is ever leaked, by the initial key distribution mechanism or by any of the parties using it, then the data encrypted by the key is no longer secure. This means that, once this key is leaked, any encrypted data that was stored in the past and any data encrypted in the future should be assumed to be compromised.

One possible improvement to the preshared symmetric key strategy is to establish unique symmetric keys for every pair of parties that wish to secure data between one another (Figure 10.2). This has the potential to limit the impact of a particular party being compromised to only the keys that party has access to. However, this results in a more complex implementation and does not solve the fundamental weaknesses inherent in using preshared keys. Also, this form of encryption does not give us any confidence that the data came from a device that is trusted. If we want that kind of assurance, and asymmetric keys are available, then the following strategies for establishing keys for encryption can yield better data security.

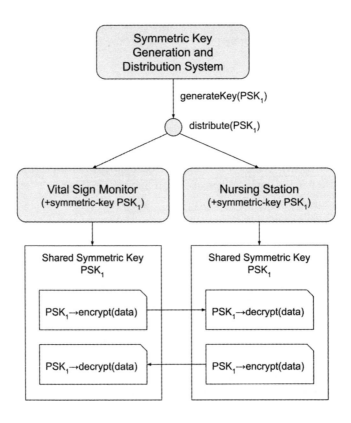

Figure 10.2   Symmetric key generation and distribution system.

10.2.1.4   Public-Key Encryption

One of those strategies is public-key encryption. If two parties that wish to communicate securely have their own asymmetric key-pairs, the parties can share their public keys with one another and use the combination of their own secret key and the other party's public key to derive a shared, symmetric secret key. This key can then be used to encrypt and decrypt data between the two parties much like the preshared symmetric key. In addition to encrypting the data, the asymmetric key pairs can also be used to create a signature to authenticate the data and provide some assurance that it actually came from the party that owns the keys.

This form of encryption also has weaknesses. If a third party were to intercept the public keys when they are being shared between the parties, then it is possible for that third party to execute a man-in-the-middle attack where the data is decrypted by the third party and read or copied prior to the data being reencrypted by the third-party and passed along to the original recipient.

10.2.1.5   Public-Key Encryption with Ephemeral Keys

Another strategy for establishing encryption keys is public-key encryption with ephemeral keys (Figure 10.3). This process is the same as public-key encryption except the public key that is used to derive the shared, symmetric secret key comes from a temporary asymmetric key-pair that is newly created for each communication session (or even more frequently). The trust of this new, temporary key-pair can be established by signing its public key with a trusted long-term key-pair, LTK, before it is shared. This form of key-agreement ensures that each session where data is secured uses a unique shared symmetric key to encrypt

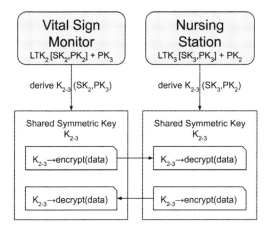

Figure 10.3   Public-key encryption.

the data. This means that any data that is captured is only compromised if the symmetric key for that particular session has been compromised. This is what is known as perfect forward secrecy (PFS) [9]. There is still the vulnerability that calls into question whether the LTK public-key that was originally shared between the parties was intercepted by a third party attempting to execute a man-in-the-middle attack. In this scenario, there is still no way to confirm that the public key belongs to a trusted party and the data could still be leaked, unless there exists a root-of-trust that can establish trust of the shared public keys.

Public key encryption with ephemeral keys (EKs), plus a root-of-trust to establish trust of each party's long-term public keys, can help prevent a man-in-the-middle attack. Given a root-of-trust, we can verify that any public key shared with us, with an accompanying certificate, is trusted. By evaluating the certificate-chain for that public key, we can determine if it has been vouched for by a higher, trusted party and that a third party has not intercepted the public key and sent their own key in order to perform a man-in-the-middle attack. Figure 10.4 shows the key-agreement process for deriving a shared, symmetric key using public key encryption with ephemeral keys.

Even with this level of confidence that the public key of another party is trusted, as always, there are weaknesses that must be understood and mitigated. If a third party is able to gain access to a higher key-pair in the root-of-trust hierarchy, then that third party can sign any key it wishes into the root-of-trust at that level, and then can perform the same nefarious acts as before.

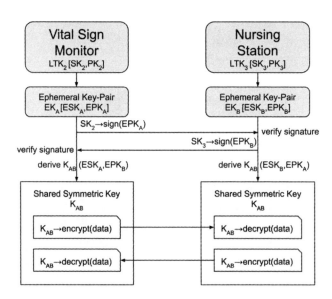

Figure 10.4 Public-key encryption with ephemeral keys.

In a root-of-trust, it is essential that the risks of key exposure are properly assessed and there is a strategy to deal with the possible leak of signing-keys that can extend the root-of-trust and fool parties into believing that data is secure and originated at a trusted source.

10.2.1.6 Key Lifecycle and Key Revocation

For a root-of-trust to properly manage the lifecycle of the keys in its hierarchy, careful thought must be given to how long keys should be valid, or trusted, and how likely it is that keys may be compromised.

Keys that are securely stored in a hardware security module (HSM) or an air-gapped network and used to extend the root-of-trust, known as signing keys, are likely to have a low risk of being compromised because security protecting those keys can be tightly controlled. For such keys, the main factor influencing how long the keys should be valid is likely the strength of the cryptography itself and how long it might take an attacker to break the cryptography with brute force, known as the work factor, and expose the secret-key.

However, keys on an asset that is connected to a public network, or is physically accessible to potentially untrusted individuals, have a much higher risk of being compromised. In order to determine how long these keys should be trusted before rekeying the asset, a somewhat subjective calculation can help inform a reasonable key validity lifetime. Other factors may also influence how practical it is for an asset to perform a rekeying process, for instance, how available a connection is to the PKI, required to approve and create certificates for the new keys.

If keys within a root-of-trust are compromised before their predetermined validity time-frame expires, then the keys may need to be actively revoked so that assets in the field know that the keys should no longer be trusted. This revocation process can be difficult to implement and maintain over time, and has the same connectivity concerns as rekeying [10].

10.2.1.7 Asymmetric Keys

There are many asymmetric key algorithms to choose from, but commonly used keys today fall into two categories. Rivest-Shamir-Adleman (RSA) keys (currently the most widely used) [11] and Elliptic Curve Cryptography (ECC) keys, which are becoming more and more popular [12]. ECC algorithms have a number of advantages over RSA keys. ECC keys are often chosen over RSA because they can achieve the same level of security with smaller key sizes and their calculations perform better in software implementations.

10.2.1.8 Symmetric Key Encryption Algorithms

The effectiveness of a particular data security solution is also dependent on the strength of the underlying cryptography algorithms. A very common symmetric key encryption algorithm in use today is known as advanced encryption standard (AES) [13]. There are a number of different modes that AES implementations can use; some are less secure than others. Many popular hardware platforms include AES hardware accelerators that make this encryption algorithm popular, especially for performance-critical applications.

A newer stream cypher for symmetric key encryption known as ChaCha20 is also gaining popularity, due to the desire to have some alternative to AES as weaknesses are discovered over time. Also, ChaCha20's resistance to timing attacks and better software-only performance are attractive.

10.2.1.9   Protecting Keys

Protecting the asymmetric secret key material in a root-of-trust, as well as the symmetric secret material, is essential to keeping data secure. Often, secret key material is kept in a key-store that encrypts it so the keys cannot be readily compromised. However, the key-store must be able to be unlocked, and the secrets decrypted, in order for them to be used. While in use, the secret keys should be kept in unencrypted storage for the shortest possible amount of time, and any memory that contained the keys should be properly and promptly cleared.

It can be difficult to keep key-stores secure since the information to unlock the key-store must be accessible to the application that needs to use the keys. This means that if attackers gain access to the device, they could find out the trick to unlocking the key-store and steal the secret keys. One way to limit the exposure of secret information is to use hardware implementations to store and lock away the keys so that they can never be read. The keys can be used exclusively within the hardware implementation to sign, verify, encrypt, and decrypt, such that there is high confidence the secret key material will remain safe even if an attacker has access to the device.

10.2.1.10   "Don't Roll Your Own Crypto"

Some software or hardware development teams might determine that selecting an open source cryptography library, figuring out how to use it, and then building and optimizing it for a given use-case, is less ideal than implementing their own cryptography functions from scratch. It's just math, right?

In reality, getting cryptography programs right can be very difficult, even at the most basic level. Aside from simply writing code that is correct and free of bugs, like memory leaks, there are plenty of things that can go wrong cryptographically when writing software functions to perform what seem to be straightforward mathematical operations. Side-channel attacks, like timing attacks, are often difficult to avoid as they require the software engineer to think in ways that are counterintuitive to their normal development practices. One hundred percent correct implementations of cryptographic algorithms, from a mathematical perspective, still yield broken cryptographic systems that leave data and/or systems vulnerable to attacks. Often these attacks are not even difficult to carry out.

There are many challenges in implementing a cryptography solution that sufficiently mitigates the risk of exploitation. Special attention must be given to timing attacks on the cryptography library itself, weaknesses in key management, misconfiguration, and the list goes on.

The preferred route for implementing cryptography in software applications is to utilize common open source libraries, like OpenSSL, LibreSSL, libsodium, and WolfSSL, amongst others, in order to leverage the countless hours of design, implementation, testing, fixes, and scrutiny that are behind them. Even with security experts writing or contributing to such libraries, there are often vulnerabilities that are discovered that must be patched in order to keep data secure. When vulnerabilities are discovered in a common open-source library, the likelihood that those vulnerabilities will be quickly patched is relatively high given the broad adoption and the large open-source communities surrounding them.

10.2.2 Applying Cryptographic Technology to Medical Devices

Cryptography is a powerful technology that MDMs can leverage to assure information confidentiality, integrity, and authenticity. However, understanding cryptography is not sufficient: medical devices require certain unique considerations.

Typical consumers of the Internet and computer systems use computer hardware and software applications that rely on cryptography to protect their computers as well as their data. When cryptography breaks down in one of these systems, the worst-case scenario is the loss of sensitive data like social security numbers, credit card numbers, and bank info. This may seem horrible, and can have far-reaching personal consequences, but does not mean that a person's life is in immediate danger. With a medical device, a failure in the cryptography it relies upon can mean personal harm or even fatality. This personal harm can happen through direct impact to the patient (e.g., wrong dosage, incorrect diagnosis) or indirectly by impacting a healthcare provider's ability to deliver care in a timely manner.

Some of the unique requirements for cryptography in medical devices include:

- Cryptographic algorithm constraints (key, signing, hashing, encryption) (e.g., resource limitations and performance requirements);

- Failure modes (e.g., when a certificate expires or a signature fails to verify);

- Certified configuration (signing the information that drives a medical device's functions);
- Secure, cross-organizational interoperability (such as an imaging system talking to a PACS system);
- Key and certificate management;
- Connectivity limitations.

10.2.2.1 Cryptographic Algorithm Constraints

Medical devices have a wide range of different hardware and capabilities. On one end of the spectrum, there are low-resource devices, such as embedded devices, that may need to be small enough to be implanted into a patient and make use of a single battery for its entire lifetime. On the other end of the spectrum, there are large capital equipment devices, with multiple, high-performance computers that are tied into a building's power system. One important thing remains constant across all of these devices—they are meant to treat or diagnose patients. Therefore, the stability and security of the devices' data and operation are essential.

The capabilities of a device's hardware can constrain how effective its data security can be. Cryptography is based in mathematics and, while computers are great at computations and math, cryptographic algorithms can be relatively demanding in terms of processing power and can only be optimized to a limited extent without compromising the security of the algorithms. Sometimes, the secret keys and root certificates that drive a device's data security need to be secured in a way that requires hardware and software capabilities that are not necessarily available in low-resource medical device hardware.

When attempting to do cryptography on small, embedded microcontrollers, processing power can be too restrictive to make certain cryptographic operations practical. These smaller devices are often designed to operate with low power consumption so that they do not drain a battery too quickly. While this is usually enough processing power to control sensors, process sensor data, and establish unencrypted communication channels, it's often not enough to perform cryptographic operations fast enough, or with acceptably low power consumption, to be practical for the use-case of the medical device. Smaller devices that do implement cryptography functions for securing data often rely on hardware acceleration to make data security efficient enough to be practical for the devices' use-cases.

On the other hand, although capital equipment that is attached to power grids may seem to have plenty of processing power to spare, there are often still challenges when introducing cryptography. Some of these larger devices process large amounts of data at high speeds. In order for the device to operate properly, it may need to process and transmit data within a very short period of time with minimal latency and high predictability. Introducing any extra calculations or processing steps, as required to perform cryptography, may cause the device to not meet its performance targets or even to malfunction. For this reason, even high-resource medical device hardware may need a boost from cryptography hardware implementations. Often, AES encryption modes are used in these scenarios, since acceleration through AES hardware implementations is more readily supported.

Regardless of the processing power available to the device, the protection of critical cryptographic secrets or configuration may require specialized hardware capabilities or crypto elements. Otherwise, critical data like secret key material and root-certificates can be compromised, causing data to be leaked and/or altered and erroneously trusted, thus opening the opportunity for an attack on the device.

10.2.2.2 Failure Modes

The handling of a failed cryptography operation must be carefully considered for each type of medical device and use case. A certain cryptographic failure mode (e.g., a failed certificate verification) in one type of device may protect that patient from harm, whereas the same failure mode in a different device type could cause harm. It is important that general-purpose cryptography libraries are able to be configured to achieve the appropriate failure modes for each type of medical device.

Two real world cases can illustrate the unique requirements for failure modes of the cryptography implementations in medical devices. In the first case, consider a linear particle accelerator, or linac, commonly used for radiation therapy. This device can destroy tumor cells without making an incision by firing a beam of radiation into a patient's body. If a patient receives the correct radiation dose at the correct location, it can destroy a malignant tumor and save a patient's life. If the patient receives too high a radiation dose and/or a dose in the wrong area of their body, it can cause irreparable harm or even death. In order to protect the patient, a linac device manufacturer might use cryptography to sign a treatment plan to ensure that it was not corrupted or maliciously changed before the machine read the plan and turned on its radiation beam. In this circumstance, we would want the device to refuse to deliver a treatment if it was unable to verify the signature on the treatment plan. This may delay the treatment, but it would also likely save the patient from being harmed by a corrupted or maliciously altered treatment plan.

In the second case, consider a pacemaker which is used to maintain a patient's heart rate to prevent cardiac arrest. The pacemaker is loaded with settings that adjust the pacemaker's operation for the particular patient's heart condition. Doctors can adjust these settings as needed using a specialized programmer. To prevent corrupted or maliciously altered settings from being applied to the pacemaker and harming the patient, a programmer can sign the settings before they are transmitted and the pacemaker can verify that the settings came from a trusted programmer. A common failure case in cryptographic signature verification is that a signature is good, mathematically, but the certificate for the key that generated that signature has expired. This can cause the pacemaker to reject the new settings. In some cases, a patient may require a settings update in order to restore their heart to a normal rhythm and their life could be in immediate danger if the pacemaker settings are not updated. Since a failure to update the settings of the pacemaker could cause immediate harm to the patient, the medical professional should have some form of override mechanism to force a settings update even if the cryptographic operation, signature verification, fails. Ideally, an alert of some kind should be propagated from the pacemaker and/ or programmer to a monitoring system to log the event for root cause analysis and remediation so that future settings updates can be performed with valid certificates and without the doctor being required to break protocol and force an update.

10.2.2.3   Certified Configuration

When a medical device performs its functions, its operation should be as deterministic as possible. The more predictably a medical device operates, the safer and more effective it is at achieving its desired outcome with the patient. From a cryptographic perspective, ensuring consistent and specified behavior can be achieved, in part, by providing the medical device with trusted, intact software and configuration data, and using trusted, well-defined security and operation parameters for its data services. This applies to internal data services, for securing data stored on a disk and other persistent storage mediums, or transport security, for safely sending data over a network.

Often, configuration parameters for a medical device are established in an insecure manner. While sometimes there is corruption protection or checksum verification built into configuration mechanisms, it is far less common that configuration data is signed cryptographically in a way that establishes high confidence that the configuration data that the device relies upon is not only intact but that it came from a trusted source. Properly secured configuration data that establishes network allowlists, security parameters for different data types, and acceptable cryptographic algorithm lists decreases the attack surface of the device and can help establish more deterministic behavior.

10.2.2.4   Secure, Cross-Organization Interoperability

Often, the subject of secure, cross-organizational data interoperability is raised in the context of medical devices that must share data, but belong to different companies (MDMs) or organizations (HDOs). For example, if a CT-scanner must send imaging data to a PACS system so that doctors can diagnose a patient, how does the doctor know that the images they are looking at originated from the correct CT-scanner and have not been altered prior to delivery? Cryptography, in the form of digital signatures and a root-of-trust, can give doctors confidence that the images they are using to diagnose a patient are intact and were produced by an appropriate, trusted medical device.

Secure, cross-organization interoperability is, of course, more difficult than constructing a data security ecosystem for a single organization and their medical devices. Incompatibilities may exist between the different types of cryptography used in the different organizations' devices.

While efforts are currently being made to address this sort of secure interoperability between medical devices, this issue has not yet been solved. Standardization of the cryptographic ecosystem and protocols surrounding medical devices may help address these challenges.

10.2.2.5 Key and Certificate Management

Managing keys and certificates to facilitate trust and data security for medical devices can present unique challenges compared to the more common PKI use-cases. Most commercial PKI platforms in the industry today lack a comprehensive solution that will work in all medical device use-cases.

Often, medical devices lack the ability to leverage PKI platforms' SDKs, written in higher-level programming languages, in order to facilitate key and certificate management operations. This may result in additional burdens on the MDM's engineering teams to implement these PKI operations on their own and figure out how to make it compatible with existing PKI infrastructure that was not designed with medical device use-cases in mind.

Existing PKI platforms may also lack support for key algorithms, or for efficient digital certificate formats, that will work in the unique hardware configurations that are often found in medical devices. A medical device manufacturer may determine that, due to hardware or performance constraints, they have a limited set of options for the digital certificate sizes or cryptographic key algorithms that will work for their device. PKI platforms may not support generating certificates that are small enough, or with the required key algorithm, resulting in no viable option for leveraging existing PKI solutions for the medical device.

Other considerations that need to be addressed when building or using a key management system include:

- Management of key lifecycle, including managing crypto period (the time span during which a specific key is authorized for use) and revocation operations.

- The complete key lifecycle is comprised of (not all are required in all cases): generation, registration, storage, distribution and installation, use, rotation, backup, escrow, recovery, revocation, suspension, and destruction.

- Reliable and secure deployment of keys during the production cycle of the device.

Proper management of key lifecycles requires a combination of a technical and procedural approach (e.g., a technical infrastructure to protect the keys and procedures to manage access to that infrastructure). The effort required will vary depending on the nature of the key: it will be high for a root-of-trust key, whereas it can be lower for a temporary key. Methods commonly applied include:

- *Procedural*: separation of duties; dual control; split knowledge;
- *Technical*: physical security; access security; user- and role-based access management; considerations for high availability and business continuity.

Detailed guidance on how to set up a PKI infrastructure and manage key lifecycles has been provided, through for example NIST SP 800-32 (Introduction to Public Key Technology and the Federal PKI Infrastructure), NIST SP 800-57 Parts 1-3 (Recommendation for Key Management), and NIST SP 800-130 (A Framework for Designing Cryptographic Key Management Systems).

10.2.2.6   Connectivity Limitations

Connectivity limitations that impact key management and the use of cryptographically secured communication may occur in a number of scenarios and use cases:

- Devices operating on a segmented or isolated network that limits connectivity with a PKI infrastructure;
- Intermittent device connectivity where a device may only be network-connected after an extended stand-alone period;
- The device may be connected via a low-bandwidth connection that may impose certain data transfer rate restrictions as it relates to the use of commercial keys or certificates;
- The device may be connected via a proprietary and/or point-to-point interface.

For security reasons, medical devices often run on segmented or isolated networks and may not have the connectivity that most existing PKI platforms require. For instance, a device may not be able to make a direct connection to hosted PKI infrastructure in order to generate its certificates and acquire certificate revocation information. In this case, extra infrastructure, or PKI gateway systems, may be required to make PKI processes work for a medical device.

Although individual devices and their respective Integration architecture vary widely, in general, the following requirements need to be addressed in the limited connectivity scenario:

- Type of key and certificate used, based on device compute power and bandwidth.

- Initial certificate deployment—typically during production process, but in unique cases other solutions may need to be developed.

- Key use during operation or communication, based on device capability and available bandwidth. Depending on use case, this could range from a simple point-to-point connection to managing connectivity in a limited subnet.

- Key rotation, revocation, or suspension. This may not be addressable, or may not need to be addressed, in all cases.

Depending on the specific device architecture and use case, these operations could be managed at time of connectivity to a network or proprietary device. For example, when managing an implantable or patient-worn device, the manufacturer could embed the device certificate (of the appropriate form factor) during production. During device operation, that certificate would then be used to assure confidentiality, integrity, and authenticity of the device communicating to an external programmer. That programmer could potentially act as a gateway in cases where the device certificate would need to be updated or revoked. However, practical and hardware limitations may outweigh the added security benefits such a process would provide, and this may not be a required feature for all device types. The specific implementation needs to be assessed on a case-by-case basis based on a risk/benefit tradeoff analysis.

In a similar scenario, device certificates could also be used to manage and assure integrity and authenticity of firmware updates deployed to the device.

In practice, there are a large number of device types, use-cases, and possible connectivity limitations that are not addressed in this high-level discussion. Hopefully, the provided examples and general discussion provide enough insight on this specific topic.

10.2.3   Available Cryptography Tools

Most commercial cybersecurity solutions on the market today are reactive in nature. They aren't necessarily designed to ensure that medical devices maintain their essential clinical performance in their deployment environment. These solutions are designed to detect issues as they arise and monitor things like network traffic or system logs from the sidelines. While firewalls and network IDS are absolutely necessary for securing hospital networks and medical devices, this is not a complete solution. Devices need to be secure by design such that they can still operate securely in a potentially hostile network environment and even a physically insecure environment.

The common data security tools on the market today that are more proactive, and can potentially provide a trust hierarchy to drive data security in the device, are deeply rooted in the world-wide-web TLS/SSL certificate paradigm where a client, such as a web browser, is meant to talk to a vast number of server endpoints, or websites, to request information to download and display for the user. Many medical devices attempt to use these same security tools to secure the data in transit between their medical devices and hosted servers.

While a medical device connecting to a hosted server, in the cloud or a data center, is similar to the scenario where a web browser is connecting to a website, there are key differences. The first and most obvious difference is that the medical device, as a security best practice, should only trust a specific subset of the servers on the Internet and not just any server from any organization on the world wide web. The subset of servers the medical device wants to communicate with is likely limited to just a few server endpoints that belong to the MDM or HDO. This type of restriction may be achieved by the MDM acquiring TLS/SSL certificates that are signed by an intermediate certificate for their organization and configuring the medical device to only trust servers that chain up to their organization's intermediate certificate. However, there are some difficulties in implementing this solution with certain medical devices. Among other issues, the PKI vendor issuing the certificates may not support an x509 format the device can handle, they may not support a certificate chain hierarchy that gives the medical device vendor confidence in the integrity of the root-of-trust, or they may not support the desired key algorithms and signing algorithms that would yield the appropriate or workable solution. Moreover, the TLS/SSL configuration likely requires special attention to ensure the device does not inadvertently negotiate down to an insecure algorithm or trust a certificate chain that is outside of its organization.

One of the more difficult aspects of implementing a PKI solution to establish a root-of-trust is the information workflow and protocol required to manage keys and establish certificates. The current certificate signing request workflows and APIs in the market, as well as the way the outstanding certificate information is stored and tracked, can be problematic for medical device systems that represent different use-cases from traditional PKI applications.

These are the issues we encounter with the simplest connected medical device use-case. Many medical devices are not this simple. Medical devices range from a single device with a simple client-server configuration, to a configuration with many computers on a local subnetwork that connect to the Internet through a gateway machine. In order to properly secure the range of medical device configurations we need tools that are specifically designed to address these use-cases. Traditional TLS/SSL certificates and their provisioning processes are not appropriate for medical devices that talk to components on a local network where traditional, Internet-centric top-level-domains and DNS resolution do not apply.

10.2.4   Cryptography in Low-Resource Devices

The previous sections discussed how cryptography can be applied to improve the security posture of medical devices. But we also need to recognize that some devices may have limitations that require a different or modified approach in order to enable cryptography in low-resource contexts, such as memory, power, connectivity, and/or bandwidth-constrained use-cases. These devices may require unique considerations so as not to compromise device functionality or useful life.

One of the most common ways the authors have seen engineering staff at MDMs incorrectly implement security into their embedded medical device is to implement risk controls that were originally designed for high-resource CPUs with unlimited memory, standard network connectivity, and no restrictions upon power consumption. In some medical devices, such as a device utilizing a mains-powered, PC-grade CPU, these big solutions can be successfully implemented. This includes leveraging PKI; however, this is generally not possible (or difficult and may require compromise) with low-resource medical devices employing a point-to-point style of communication.

Many of the medical devices being designed today are body-worn or implanted, which results in physically small devices with low-capacity batteries and power-conserving microcontrollers (MCUs) being utilized in these designs. Ten years ago, this type of MCU would have very limited resources, including CPU throughput, data RAM space, program flash memory space, and poor support for low-power modes of operation.

The current MCU design, however, is a completely different animal. Newer MCUs are designed for extremely low-power modes of operation, have much larger internal RAM and program memory sizes, higher throughput, and most importantly (for the topic of this book) dedicated security functionality implemented directly in the core of these MCUs. While still low-performing compared to their PC cousins, newer MCUs are far more capable than previous models in this class.

Security features that can be found in many new MCUs are:

• True random number generator (TRNG);

- Hardware-accelerated cryptographic engine (AES, ECDF, SHA, HMAC, CMAC, RSA, ECC);
- Secure boot with a hardware root of trust;
- ARM® TrustZone® virtualized CPU for trusted executables;
- Locking and unlocking of debug port;
- Differential power analysis mitigations;
- Antitamper mitigations;
- Physically unique function (PUF) for each MCU;
- Secure key management and storage;
- Support for secure firmware updates.

This new set of security features permits engineers to create stronger security controls with good design practices, and still have minimal impact upon the rest of the MCU's resources (e.g., power consumption) and on device performance, availability, and longevity. The following sections will give examples of how these tools can be used to secure a medical device system.

10.2.4.1   Implementing Key Exchange in Low-Resource Devices

When performing encryption (or any other cryptographic operation that requires a key), the weakest part of the process is the key. All aspects of a key's lifecycle are a source of vulnerabilities, including key creation, sharing, storage, and modification or deletion.

Key creation

Medical devices do have an inherent advantage over commercial products, such as web browsers or apps, that are nonunique in nature and so require that a key be exchanged dynamically, via methods such as ECDF [12], RSA [16], or one of the many variations of encrypted key exchange [17]. Medical devices have been controlled during manufacture, which allows the MDM to inject an initial seed key that can be used to derive other subkeys using a key derivation function (KDF) for instantiated uses (the same key should never be reused between any of the various possible cryptographic utilizations). This injected seed key should be unique to each device and stored securely on the device. During manufacturing the injected key should be associated with a publicly available device identifier, such as a serial number.

An even better approach is to utilize a PUF which is already present in the MCU and unique to each MCU, thus removing the need to inject a key during manufacturing.

Key sharing

The approach of keys being preshared at the time of manufacturing has already been discussed, but there are other ways that key material can be shared, each of which varies according to use case and strength of protection.

If the two devices have human interfaces that allow for displaying information and a method for the user to input information, then this can be utilized as an out of band (OOB) method of sharing a key via numbers or even as a phrase of randomized text.

OOB gives us many solutions to share a key, including utilizing a devices' LED to blink out a serial stream of data to a camera on a smartphone [18], or the device producing a high-frequency (above human hearing) modulated data stream to be received by a smartphone's microphone [19]. Another OOB approach is for the device to have a 2D barcode label that can impart an authentication key to a smartphone via its camera, which in turn can be used by the smartphone app to allow it to connect to the device via a wireless connection. All of these very specific use cases employing OOB solutions depend on physical proximity to the device to convey trust.

Key storage

Storing a key can be problematic as there are physical attacks that could expose these keys. However, there are separate electronic components called secure elements that can be used to securely store key values. Some newer MCUs offer secure internal memory or a separate, virtualized security CPU, such as Trust-Zone, which can be used to store keys in the MCU's memory.

Even when utilizing old MCUs, if the key values are unique to a device (as they should be) or change rapidly (ephemeral), then exposure of one of these keys has a very limited impact due to its short life.

Key deletion or modification

The respective methods used for created or shared keys impacts end-of-key viability, as there is no certificate authority to delete a key. This presents a challenge unless key use is coordinated by a central authority, such as the MDM's key-to-serial number database.

Another approach to key use is to make the viable lifespan of any key utilized for encrypting communications very short (ephemeral), as in perfect forward secrecy. This results in constantly generating new keys for new messages being transmitted, so exposure of a current key can at most only expose a limited amount of data (e.g., the device's latest status message). There have been multiple schemes created to achieve this system, such as using a Diffie-Hellman exchange for each key change, but probably the best-known and most widely used is the Signal messaging app, which employs the double ratchet protocol invented by Moxie Marlinspike and Trevor Perrin [20].

At the most fundamental level, Signal creates a new key for each message being transmitted, and this is performed on both the sending and the receiving devices through the use of predictable one-way functions (in the case of Signal, a KDF). While Signal is an extremely secure messaging app that can stand up to nation-state level attacks [21], the basic concept of perfect forward secrecy can be implemented in a number of ways, including leveraging SHA, HMAC, or CMAC hardware accelerators to rapidly hash the existing key value to create a hashtag that will be the next key value. This easy-to-implement approach makes for a very secure system; however, the challenging part of such a system is recovering from a loss of key synchronization between the sender and the receiver in a manner that cannot be exploited. It should also be pointed out that the encryption key used for one direction of communications should not be the same key used for the return direction and both should be ephemeral.

The observant reader may have recognized that there are multiple possible variations upon this theme, such as expanding the PUF value (by the use of a KDF) into a set of function-specific root keys, so that each root key can also be ephemeral as shown in Figure 10.5.

While such approaches are typically used for data in motion, it can also be extended to create a data-at-rest encryption scheme as well.

10.2.4.2   Implementing Encryption in Low-Resource Devices

As discussed before, encryption is a method to convert a data set (plaintext) into another data set (ciphertext) that appears to be random data, and thus cannot expose the contents of the original plaintext. Performing this transformation requires an encryption key; decrypting the ciphertext back to the original plaintext requires a decryption key.

There are many different types of encryption, but the two major categories are symmetric encryption, where the encryption key and the decryption key are the same value, and asymmetric encryption, where the encryption key and the decryption key are different but associated values referred to as a key pair. Individually, these keys are referred to as the public key and the private key.

Asymmetric encryption is used on the internet due to the ease of key management, however, it is not better or more effective encryption than symmetric. In many ways, asymmetric encryption is worse than symmetric, because asymmetric is thousands of times slower than symmetric and the length of key required to produce an equivalent strength of encryption is much larger with asymmetric encryption. NIST has calculated that a symmetric AES key of 128 bits in length as equivalent to an asymmetric RSA key of 3072 bit in length, and a symmetric AES key of 256 bits in length to an asymmetric RSA key of 15360 bits in length [22]. Therefore, in almost all cases for utilizing encryption in medical devices, the best choice is symmetric.

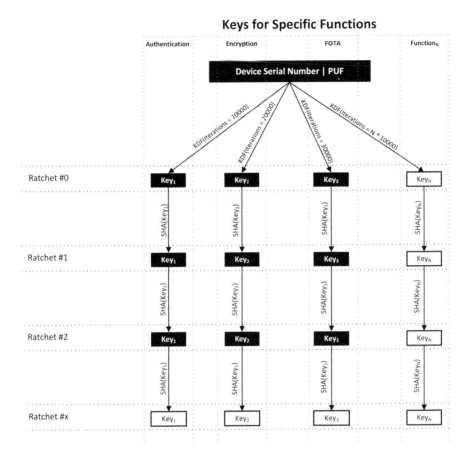

Figure 10.5 PUF value expansion using function-specific ephemeral root keys.

Advanced encryption standard (AES) is the most likely form of symmetric encryption to be supported by MCU hardware acceleration. AES is frequently available in 128- and 256-bit lengths. Both are secure at the time of this writing, however, due to the long operating life of a medical device (up to 20 years) and the probable improvements to hardware that might be used against the device during its life, wherever possible 256-bit encryption should be used.

When selecting an AES mode, hardware acceleration support may again be the deciding factor—not all modes are supported by these accelerators. Each mode has its positives and negatives (shown in Table 10.2) that have to be weighed for each use case, but generally speaking, GCM mode is the mode that should be preferred. It has the fewest downsides and supports integral hashing of the ciphertext to provide not only confidentiality but integrity checking as well.

Table 10.2
Supported Functionality of AES Modes

Mode	Name	Authenti-cating	None required	Padding	Pseudo stream cipher	Storage Encryption	Self-synchro-nizing	Chosen plaintext vulnerable
ECB	Electronic code book		✓					✓
CBC	Cipher block chaining		✓					✓
CTR	Counter				✓			
OFB	Output feedback				✓			✓
CFB	Cipher feedback				✓		✓	
XTS	XEX tweakable stealing	✓				✓		
CCM	Counter cipher message	✓						
OCB	Offset codebook	✓	✓					
GCM	Galois counter mode	✓						

10.2.4.3   Implementing Authentication in Low-Resource Devices

One of the single most important security controls that can be implemented in a medical device system is authentication, both human-to-machine and machine-to-machine. Human-to-machine authentication has been in use for a much longer period of time and, as a result, the technical methods are more widely known, such as account credential, password, and a salt value concatenated together and then hashed with SHA-256.

For machine-to-machine authentication, the methods are completely different. On the internet, we can leverage PKI (via x.509 certificates) to achieve authentication, but that requires the presence of a trusted third party, the certificate authority (CA). In many medical devices, the connections to be authenticated are point-to-point connections without the possibility of a trusted third party. In these cases, a separate level of authentication needs to be performed between the two connecting devices and this needs to be performed in a manner that:

- Cannot be replayed (a replay attack is where a valid authentication is recorded by an attacker and then replayed back without understanding the values or algorithms being utilized);

- Does not expose any shared secrets to a man-in-the-middle attack against multiple iterations of authentication;
- Can be performed immediately following a connection being established before any other medical device functionality can be performed;
- Uses a unique value for each medical device so that a large-scale attack cannot be derived from a successful attack on one device.

The usual solution to this is a challenge-response process that is tailored to the limited resources of an MCU. Specifically, it leverages the one-way function of a keyed cryptographic hash (i.e., an HMAC) to demonstrate knowledge of a shared secret (in this case the key). In the sequence diagram (Figure 10.6), Alice contributes half of the random value to Bob to prevent the possibility of a

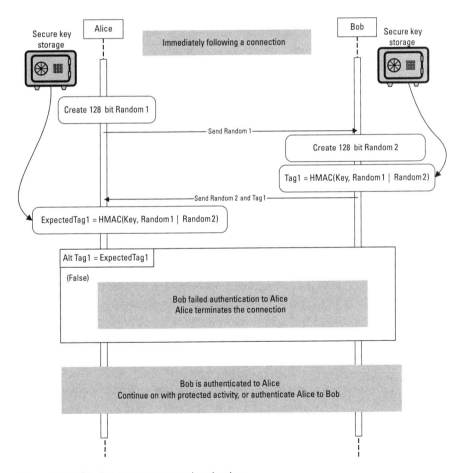

Figure 10.6 Challenge-response authentication.

replay attack if Bob supplied all of the random value. The key value should be unique for each Alice, as discussed in the section about key creation.

10.2.4.4   Implementing Integrity in Low-Resource Devices

In the domain of medical device security, integrity can take on several meanings, such as integrity of the intended function or integrity of data at rest/in motion. For the purposes of this section, we are only addressing data integrity.

Traditional data integrity tools consisted of algorithms such as a CRC, checksums, and parity bits. However, these are not cryptographically significant, so they need to be replaced with cryptographic hashes.

There are three fundamental ways to protect data: (1) by signing it, (2) by adding an HMAC/CMAC to the data, or (3) by utilizing a mode of encryption that conveys integrity as well as obfuscating the data (such as AES-GCM mode).

Signing is typically utilized in a system with asymmetric encryption. This is where the data set is hashed with a nonkeyed hash, such as SHA-256. The process is as follows:

- The data set is encrypted with the sender's private key;
- The ciphertext is then hashed with a SHA-256 to create a hashtag value;
- The hashtag value is appended to the data set and then encrypted with the public key of the recipient;
- The recipient then uses their private key to decrypt the ciphertext;
- Recipient validates the SHA hashtag value;
- If the integrity is intact, then the recipient decrypts the data set back into plaintext using the sender's public key.

One way to think of this signing process is that the authenticity of the sender is conferred by the asymmetric encryption activity, while the integrity is assured by the hashtag value. This is usually performed in the context of asymmetric encryption, but it doesn't have to be: the data set can be concatenated to the hashtag value and then encrypted with symmetric encryption. This is not as flexible or secure as when implemented in an asymmetric context, but it is still a valid approach.

Directly utilizing an HMAC or a CMAC (both keyed hashes) is also a valid method to ensure the integrity of the data set. This is a symmetric approach to integrity, where both the sender and the receiver have previously established a shared secret in the form of a hash key. The data set is subjected to an HMAC function with the previously shared key and the resultant hashtag value is appended to the plaintext of the data set being transmitted. The receiver vets the integrity of the plaintext by performing the same HMAC operation and comparing it to the received hashtag value. Usually, after the sender appends the hashtag value to the data set, the entire structure is then encrypted (asymmetric or symmetric) before being sent to the recipient.

10.2.5 Cryptography—Conclusion

Cryptography can provide for very strong protections against many attacks and exploits but requires careful selection of methodology, process, and hardware (e.g., the selection of the right algorithm, key lifecycle scheme, and correct MCU) early on in the design lifecycle of the device. Cryptography can be used effectively to protect data confidentiality (encryption), integrity (hashing), and authenticity (signing). Considering resource restraints of many medical devices and use cases, not all are required in all cases (e.g., why spend cycles in encryption if only integrity or authenticity are required?).

But technology is always on the move. As hardware improves, asymmetric encryption solutions may become possible, even in ultra-low-resource contexts, as both built-in and stand-alone hardware-accelerated implementations of ECC and other cryptographic technologies become more common and begin to close what is, as of this writing, an unacceptable performance gap for most embedded medical device use-cases.

There are many more ways cryptographic primitives can be used to protect a medical device than we can address in this book, and many sources of additional information about how to apply them in a design. Foremost among these is Bruce Schneier's excellent book, *Cryptography Engineering* [1].

10.3 Securing Communication Mediums

10.3.1 Bluetooth Low Energy

Over the last several decades of medical device development, there has always been some communication medium currently in vogue. The RS-232 serial port with a DB25 connector was replaced by a DB9 connector, which was succeeded by USB, which mutated to USB OTG. Following that, we had a series of wireless mediums (such as Zigbee, IRDA, proprietary protocols, etc.).

Then, beginning in 2010 with the ratification of the Bluetooth Core Specification v4.0, Bluetooth low energy was added into the Bluetooth universe along with the original form of Bluetooth BR/EDR. (It should be noted that BLE and Bluetooth BR/EDR are very different in just about every way, however, they are both defined and maintained by the Bluetooth SIG).

We are not going to dive into the technical details as there are many websites that discuss various technical aspects of BLE, but the definitive text is the Bluetooth SIG's Bluetooth core specification [23]. However, it can be difficult to read and locate specific topics in this very large document. A much more readable source of technical information is Mohammad Afaneh's free ebook, *Introduction to Bluetooth Low Energy* [24].

10.3.1.1 Legacy BLE Specifications

Lesson one: Don't use BLE v4.0 or v4.1 for anything. The key exchange process is badly flawed, allowing a paired/bonded BLE connection to be taken over by an attacker. Lots of online documentation exists describing the vulnerability, as well as easily downloadable exploit software. Moreover, data length extension (DLE) isn't supported until BLE v4.2; DLE is desirable because it increases the BLE payload size from 27 bytes to 251 bytes. This is useful for cryptographic operations, which typically require large numbers, as well as faster firmware updates and log downloads, and increasing overall BLE throughput from approximately 6 Kbits/s to 270 Kbits/s. (You can get even more throughput by using BLE v 5.0 or greater; the 2 Mbps link speed can push it up to 1400 Kbits/s, assuming embedded BLE radios are used for both sides of the connection [25].)

Use of BLE versions older than 4.2 indicate an older BLE radio, which probably costs more than a new BLE radio, is closer to EOL, and won't support many of the other BLE features the device may need to leverage. Therefore, any new medical device should be designed based on BLE v4.2 or above to select secure connections only mode (meaning it is forced to security mode 1 level 4, as recommended by NIST) [26]. This is as secure as you can make standard BLE over the air (OTA) security.

Unfortunately, this isn't good enough. There have been many published attacks [27] against the BLE stack, and while using legacy connections is much worse than v4.3's secure connection (which utilizes ECDH), both have had successful attacks against them. The most recent attack on BLE as of this writing, dubbed SweynTooth [28], yielded 12 new vulnerabilities as shown in Table 10.3.

These vendors' components are deployed in a large collection of devices, including medical devices from major MDMs, not all of whom are identified in the disclosure. This is also a good example of the challenges of vulnerability communication. What is the best way for an HDO to receive appropriate vulnerability information on underlying technology that is used in a specific device, but may or may not affect them? Just because the device runs a particular software does not provide enough information about the actual vulnerability to the device.

It should be also noted that, due to its relatively low current draw, the Dialog BLE radio is a popular choice for implantable medical devices, where it is frequently used not just as a BLE interface device but also as the host MCU in the medical device. In this type of deployment, a crash or deadlock doesn't just shut down BLE communications but also causes the essential performance of the medical device to cease as well.

Table 10.3
SweynTooth Vulnerabilities

Result	Vulnerability	BLE Vendor	CVE
Crash	Link layer length overflow	NXP Cyprus	CVE-2019-17519 CVE-2019-16336
	Truncated L2CAP	Dialog	CVE-2019-17517
	Silent length overflow	Dialog	CVE-2019-17518
	Public key crash	TI	CVE-2019-17520
	Invalid L2CAP fragment	Microchip	CVE-2019-19195
	Key size overflow	Telink	CVE-2019-19196
Deadlock	LLID deadlock	NXP Cyprus	CVE-2019-17060 CVE-2019-17061
	Sequential ATT deadlock	STMicroeletronics	CVE-2019-19192
	Invalid connection request	TI	CVE-2019-19193
Defeats secure connections	Zero LTK installation	Telink	CVE-2019-19194

10.3.1.2   Securing BLE

Not discussed in the SweynTooth disclosure mentioned above, but nevertheless a vulnerability in the design of some devices, is that when a BLE radio's CPU is utilized to perform the medical device's essential performance concurrent with BLE communications, an inherent conflict can occur. In all BLE SOC radios, when a BLE event (reception or transmission of a BLE packet) happens, the MCU in the BLE radio interrupts the execution of the application-layer software to process the BLE event in the BLE stack. This means that for milliseconds the main application stops functioning, along with whatever medical device functionality it was supposed to perform. Therefore, a simple attack is to pair to the medical device where the minimum connection interval is set to 4,000 ms (as an example) and then request a change to the connection parameters to a minimum connection parameter of 7.5 ms (lowest possible setting defined in BLE). This leaves the MCU with very little time to perform the intended application, potentially too little. As mitigation, the application should be tested under stress with the connection period at 7.5 ms, or the BLE stack should be configured to never allow the minimum connection period to drop below a predetermined working period.

It should be noted that as BLE is typically defined, the central BLE radio (typically a smartphone) is the ultimate authority on the connection period; however, this just means that the attacking BLE connection will terminate, as the central BLE will be set to 7.5 ms while the medical device BLE is still at 4,000 ms (or some lower threshold significantly above 7.5 ms).

This clearly demonstrates that trust cannot be conveyed by means of standard OTA security as provided by BLE. Furthermore, all the OTA weakness examples above assumed that this topology was an embedded BLE radio communicating to another embedded BLE radio. If the topology was instead an embedded BLE radio communicating to a smartphone, then it doesn't matter what OTA security exists, because the data stream between the BLE radio in the device and the smartphone app is in plaintext and has no protection applied to it (confidentiality, integrity, or authenticity protection, as discussed earlier in this chapter). Any app (or malware) on the smartphone (Android and iOS) can sniff and inject data from/to that data stream.

This indicates that for a medical device, a second layer of BLE security controls must be put in place. The steps to be followed must be performed in order (refer to earlier sections in this chapter for discussion of each cryptographic technique):

1. Authentication between the app and the medical device (Are you the correct app?);

2. Authorization between the app and the medical device (What role are you connecting as?);

3. Key exchange between the app and the medical device (the key used for encryption);

4. Enable encryption between the app and the medical device;

5. Depending upon the type of encryption being utilized, it may or may not be necessary to support integrity checking of the BLE payloads with an additional key.

There are many variations on this approach. For example, in step 3, the app could connect to a server to reference a unique key to use for key exchange. Likewise, the app may never have any knowledge of the key, and the encrypted payload would go to a server for decrypting and then selective plaintext would be returned to the app via a secure TLS connection.

Another approach is that when the key is exchanged it would then simply serve as a seed key to use in a highly ephemeral key structure, where the key would change every packet or every minute, and so on. These types of structures are one of the most secure approaches, seen in perfect forward secrecy [29] or the double-ratchet algorithm, which can serve as a basis for lightweight embedded solutions where cryptographic accelerators in hardware can be utilized.

One last point about BLE is that while there have been cheap BLE sniffers available for years, these were low-level tools that took a lot of effort to yield any results. Good high-end tools cost approximately $10k, but that is changing. There is a new low-cost SDR-based BLE sniffer called Quince [30] that should be on the market soon.

10.3.2   Other Mediums

Other frequently seen, traditional communication mediums in medical devices include serial, USB, Inductive, and MedRadio communications. (Other mediums include ethernet, 802.11x Wi-Fi, and cellular baseband.) These still-popular legacy mediums all share the common trait of being highly insecure specifications, coming from a time when most MDMs didn't consider security. Thus, when you find these types of mediums in use, the medical devices incorporating them generally have few or no security features at all.

No communication medium is secure by nature. It is only the capacity for additional layers of security to be enabled inside of the medium that matters. So, starting with the basics, and as previously discussed, the system must be able to perform:

1. Authentication;
2. Authorization;
3. Key exchange;
4. Enable encryption;
5. Depending upon the type of encryption being utilized, it may or may not be necessary to support integrity checking of the payloads with another key (via a hash function).

The five steps above should be utilized on any connections carrying PHI, PII, or command-and-control capability.

10.3.2.1   USB

USB is omnipresent in our digital universe, however, like the other mediums presented in this section, it can be easily and cheaply sniffed and injected into (as in man-in-the-middle attacks). Therefore, as in any communications, the five steps above need to be performed to protect this medium's communications.

This medium also has a unique weakness. When it initially connects to a commercial operating system (Windows, Mac, Linux, etc.) it can identify itself as a whole host of various devices, including: a human interface device (HID) like a keyboard or mouse, a mass storage device such as a flash drive, or a network interface. Furthermore, a USB device can masquerade as all of these things at the same time, permitting the following attacks against a commercial OS:

• It can pretend to be a keyboard and type in prewritten commands;

• It can inject malware into your system via its mass storage connection (e.g., simply by utilizing the OS's autorun feature) (many medical device security incidents of this type have been reported);

- It can act as a network interface and open a connection to a compromised or malicious website hosting a drive-by infection package.

This is why it is very difficult to harden a commercial operating system platform when the USB is not disabled. CERT has provided guideline addressing this [31]. Also, many commercial security products (antimalware or HIDS/HIPS, discussed earlier in this chapter) include security features specifically designed to close this USB-based attack vector on systems with commercial operating systems.

10.3.2.2   Inductive and MedRadio

Inductive and MedRadio are frequently seen in implanted medical devices as they communicate well through human tissue (unlike BLE, which needs more power to propagate through tissue due to its high operating frequency of 2.4 GHz). In implantable devices, the most important design requirement is complying with very restricted battery life constraints, due to the physically small space in such devices and the difficulty of replacing batteries or devices (as that would require a surgical procedure in most cases).

With either of these mediums, one attack vector is to just keep waking up the implant repeatedly. This consumes much more battery life than typical usage (which might be once every few months in a doctor's office), leading to early battery depletion and the need for surgical remediation to replace the implant, and this comes with all the usual risks that accompany surgery, as well as associated costs.

In the case of MedRadio, there is a 2.4 GHz wake up signal that can be addressed to a specific type of device, or just a general "wake up anyone who can hear me" signal. In either case, there is no built-in security as the specific wake-up call is in plaintext and can easily be sniffed utilizing an SDR and replayed. Or, the attack can simply utilize the general wake-up call, which has an 8 bit company ID field with a default value of 0x01. All that is needed to attack this company ID is to just run through all possible 255 values to wake the radio up. For an attack such as this there is no practical attack range, meaning that a whole house, HDO, or neighborhood could be subjected to a high-power RF wake-up call every few seconds. If possible, limit the amount of time the MedRadio stays aware, and possibly even create a back-off period where the radio cannot be woken for a set period after its previous wake event. Also note that MedRadio's main data channels do not support any native OTA encryption, so encryption must be performed at the application layer.

In the case of inductive communications, there are common myths held by many MDMs (see Chapter 11), such as where each MDM has created its own protocol and they believe that this somehow makes it secure, when it in fact does not [32].

There are also misconceptions about the maximum communication distance (usually very short) acting as a security mitigation. In reality, communication distance is a function of coil diameter, so instead of a wand-style physical format with a range of centimeters, a briefcase-sized coil could be used to attack over a greater distance (from across the building, down the street, etc.).

With any of these legacy protocols, follow the five steps using good strong cryptographic primitives to create a secure medium.

10.3.3   Hardware and Physical Interface Security

This is another one of the many areas where an embedded device differs from the security used in IT/MIS environments. Unauthorized persons can be prevented from walking freely around a server farm, yet an embedded device can be easily procured by a security researcher (or potential adversary) who could then analyze the device's vulnerabilities at their leisure. No chance of communicating logs back to servers where intrusion detection could take place, in fact, no awareness by anyone (except the attacker) that an attack is in progress.

The nature of attacks at the physical level are not as widely discussed, so frequently MDMs downplay the possibility of them occurring or ignore their potential as they may be considered out of scope and the emphasis on cybersecurity purely from a network perspective results in blindness to a large part of the device's attack surface. In fact, these are routinely used, and with devastating effect. So, let's walk through a playbook of how a physical attack could proceed.

First, look at the device. What obvious conclusions can you reach by examining the device? Are there wired communications connectors, like RJ11 Ethernet, USB, or DB9 serial ports? If so, these are your first area to look for vulnerabilities to attack. Plug them in, see what you get, perform communications at all the common baud rates, perform a port scan, see if they left the telnet service running. Nonstandard connectors may be present—they just take a little more work after the enclosure has been opened.

Look online at the MDM website. What do they say about maintenance and servicing? Is there some clue about utilizing these ports? Go look at older similar devices from this MDM. Look for more details about usage, communications, and service credentials. Try these older credentials on your unit under attack. It is common for MDMs to continually reuse credentials that were defined 40 years ago, like admin/123456 (some attempt to justify this practice by saying that the field service reps would not be able to remember new ones).

Is there some downloadable executable? If so, run it and sniff the communications. Is there any form of authentication or authorization? Is anything being transmitted in plaintext (e.g., DICOM, HL7, or plain ASCII)? Are there any obvious key exchanges taking place? If you have multiple units that can be attacked, try this on several units and analyze the data for any differences in the communications that might indicate unique authentication at work. Did the executable ask for any unique data items like the unit's serial number? This can point to the algorithm used for unique authentication.

Go after the executable. Most likely it is written in a managed language; if so, use a decompiler to return it back to human-readable code. What you are looking for are hardcoded credentials, packet structures, such as a CONST data structure of commands. Are there holes in the commands? Try those values. What about command values above and below the value in the CONST data structure? Anything else that looks interesting, like hidden functionality accessed by clicking on a specific blank part of the screen? (Amazingly we have seen this done more than once!)

Look at the device's technical documentation (including labeling and manuals). Is there an FCC ID number on it, indicating a wireless RF communications medium? Then search the FCC's website for the background technical documentation, such as band or frequency of operation, type of modulation, even picture of the PCBA. Identify large active components from those pictures. Start to build up your knowledge about the electronics being utilized, download datasheets from the component manufacturers, and especially errata sheets from them as well. While you're there, explore any development tools they may have available, or whitepapers on security or firmware updating. Perform general searches of the internet for known attacks against any of the large components you identified in the pictures.

If the device has a standard wireless interface, start attacking that by connecting to it, fuzzing the interface, performing port scans, and so on.

All of that, and we haven't even opened the enclosure yet! In fact, you may not need to, you may already be in.

Ok, we have squeezed this untouched device for about as much as we can get from it. Let's continue our quest by delving deeper into the device. Remove the enclosure. Sometimes this is a simple as using a flat blade/Phillips/Allen screwdriver, but sometimes the MDMs thought they were being tricky and utilized secure screws. This hardware is anything but secure: for $13 you can purchase a kit to defeat almost all of them (there are some very small screws utilized in smartphones that require a different kit to access). Or maybe the MDM wanted the device to really be sealed—water and dustproof—so they ultrasonically welded the enclosure together, or even glued it together. Then you want to reach for your handy Dremel rotary tool or a zero kerf pull saw or cheaper equivalent. These can be used to slice open the case at the obvious parting line of the enclosure (don't just go hacking the enclosure up, stay to the parting line). The pull saw is best: it is slower and has less chance of damaging the PCBA(s) inside the enclosure. Or did the MDM put a lock on the enclosure? Well, you can drill it out (makes a mess of metal flakes inside the enclosure), but you may realize that the MDM didn't use unique keys for each unit! It is just a matter of identifying the correct common key that was utilized, and these are always available to the public, such as the A126 key or the CH751 key. A $4 key isn't much of a security control.

Once inside the enclosure, carefully remove the PCBA(s) and start documenting everything. Note the laser branding on the large components and PCBA-mounted connectors and test points. The silkscreen labels on the PCBA can be a big help as well.

The connectors could be a JTAG or SWD port, or a serial port. Both of these would need additional investigation into them to see if they were left active. If you can use a JTAG or SWD port to extract the binary contents of the program memory from the MCU, then we have hit the hacker's jackpot!

Did that serial connection appear to display some data during power-up of the device (that's why you don't want to damage the PCBA during the removal of the enclosure)? Find the correct baud rate and read what is being output on that port. You are trying to identify the boot loader. Also, check the semiconductor manufacturer's website to see if there is a boot loader that can be downloaded.

A common boot loader is UBOOT; if this is the boot loader being utilized, try the MD command [33] to drop the contents to the display and capture it. Parse the captured information back into a binary format and decompile this binary file back to readable C code [34]. (Note that this is not the only thing you can do at UBOOT console. Colin O'Flynn has an excellent writeup on exploiting UBOOT [35].)

Returning to the PCBA, examine the components to see if a nonvolatile memory is being utilized. This might be mapped to program space or contain other interesting values.

Continuing to go deeper, the next step is to attempt to decap the MCU with the goal of exposing ROM, Flash, or JTAG fuse (so JTAG can be reenabled) [36]. Granted, this isn't as critical of a vulnerability as it used to be. With the increased use of small geometries on the die, any die below about 90 nm means the equipment to perform micro probing becomes very expensive, maybe not practical for a malicious adversary but certainly in the realm of the possible for a nation state actor that is after intellectual property. However, as MCUs shrink to 40 nm and below, flash memory can no longer be on the die, thus creating a different vulnerability.

An MCU side-channel attack is an attack that exploits unintended information exposure to infer additional sensitive information exposure, such as cryptographic keys or shared secrets. The specific unintended information exposure comes from unexpected areas, such as the amount of current being drawn by an MCU [37], the RF emanations coming from an MCU, or timing differences. All of these occur while some crucial activity is being performed by the MCU. Typical activities to monitor are the comparison of multibyte values, moving of encryption keys in memory, the encryption process, any values being manipulated in a predictable bit order. With these side-channel attacks, the goal is to expose the secret values, bit by bit. To avoid timing attacks, decisions for critical values should have operations on those values take constant time, no matter the value of the content.

For power monitoring, an oscilloscope can sometimes be utilized, but special purpose hardware exists to make the attack easier and quicker to perform [39]. These types of attacks have yielded some impressive results [37, 40]. As indicated earlier, any predictable pattern can be exploited in a side-channel attack, so randomization can be an effective mitigation [41].

10.3.3.1   Hardware Considerations

As recently as 10 years ago, finding an MCU that had any hardware-based security tools or protections was a rare event—fortunately, this is no longer the case. Besides being faster than pure firmware alternatives, and consuming less RAM and program space, there is no better anchor for your set of security mitigations than a hardware root of trust. This is true even considering the occasional hardware vulnerability being disclosed; it is still the more reliable and most secure approach.

Some of the more useful tools at the security professional's disposal include:

- *Secure boot*: Confirms the integrity of some/all of the executable program space before transferring execution to it. This can be the base of a chain of integrity checks, where the boot code is checked and then when it is executed the operating system is checked, and then it, in turn, checks the application code.

- *Cryptographic engine accelerator:* A set of cryptographic functions typically including symmetric and/or asymmetric encryption algorithms of various bit widths, as well as keyed and unkeyed cryptographic hash engines.

- *True random number:* A very useful tool for creating values for authentications and unique keys.

- *Memory management unit:* Offers the ability to map the addressable memory space for certain specific functionality, linking executable, read-only data, and so on.

- *Side-channel protection:* Adds uncertainty into execution timing and power draw to help obfuscate internal operations.

- *Physically unclonable function (PUF):* Each MCU has its own unique value that never changes, ranging in size from 64 bits to 128 bits in width. These can be useful for uniquely addressing a specific device or acting as a seed in other identification schemes.

- *Secured debugging ports:* Prevents JTAG/SWD connections to the MCU to extract the contents of internal program memory for reverse engineering. These are available in a fairly wide range of options, from protecting read, to protecting write, disable (reenabling it erases flash), and reenable based on hashtag value of the internal program space vector table.

- *Die geometry:* A nice side effect of semiconductor manufacturers seeking smaller and smaller die geometries is that as you pass the 90-nm size, the die gets very difficult (and expensive) to microprobe. However, as we reach 40-nm geometries, we will no longer be able to have flash memory on the same die, which creates vulnerabilities for external flash memories.

- *Address randomization:* A few MCUs have a feature where external flash program space is accessed in a random fashion, with the key generated uniquely for each MCU. This means the contents of the external flash cannot be mass programmed but needs to be programmed via an interface into the MCU. This does make it difficult to easily extract the executable for the external memory.

- *Virtualized-secure CPU:* Some MCUs (such as ARM cores and Trust-ZoneM) are now supporting an isolation technique where more-trusted code can be run completely isolated from less-trusted code. This can be used to contain cryptographic keys, as well as to perform dynamic integrity checks of the program space code, to prevent injection attacks.

All of these features should be considered and leveraged in your device to the fullest practical extent.

10.4   Network Security

Although in most cases network security is part of the operators' responsibility (HDO or contracted service provider), MDMs need to understand best practices to (a) be able to design devices that can be operated on a secure network, and (b) provide a system of systems that use networks as part of their functionality.

By no means, though, should an assumption about the device operating on a secured network be used as a substitute for securing the device itself—whether it is the security baseline designed into the device or as long-term mitigation in response to a newly discovered vulnerability. Although the device network is part of an organization's overall defense in depth approach, it cannot and should not be used as the primary measure to secure the device. Another way of looking at it is that network security's primary purpose is to secure the network and protect the devices on the network, but not to secure the device.

Typical elements and components of network security are network architecture, firewalls, IDS and IPS, honeypots, and DLP.

10.4.1   Network Architecture

The commercial network security tools discussed in the remainder of this chapter are complementary to a secure network architecture. A technical discussion of this would exceed the objectives of this book and has been covered elsewhere. It should be stressed, though, that proper network segmentation is an important aspect of any security architecture and provides an additional (but not perfect) layer of security. Good network segmentation does improve protection against network-based attack vectors but does not protect against air-gap attacks, for example via a data carrier. But in either case, it does help with the containment of an attack or outbreak (i.e., it provides a degree of damage control).

Decisions on how to segment a network vary and may be done based on applying multiple criteria:

- *Functional integration*: Segmentation should support and not hinder the data flow between systems and any complex integration across network boundaries, although not totally avoidable, should be minimized.

- *Organizational structure*: May align, to an extent, with the functional integration needs and helps to align security responsibilities with organizational responsibilities.

- *Geographic distribution*: Segmentation based on location (building, floor, on/off campus, etc.).

- *Risk and value*: Higher-risk and high-value systems should be cordoned off from main networks, providing higher levels of protection for higher-risk items.

- *Redundancy*: With segmented networks providing containment of a potential attack or outbreak, devices supporting the same clinical function should be segmented off so to assure that partial functionality is maintained. For example, if a hospital has six Cath labs it may be advisable to break them up into two network segments of three each to avoid total shutdown of interventional imaging services in case of a cyberattack (e.g., a malware worm).

In reality, the final decision will be a balance between the criteria suggested above and practical considerations. In fact, there may be conflicting priorities between, for example, functional integration and the need for redundancy.

A comprehensive and well-guided risk assessment approach is advisable and is, for example, described in the U.S. Department of Veterans Affairs (VA) "Medical Device Isolation Architecture Guide" [42]. Although dated from a network technology perspective, it does outline a viable decision process for risk-based network segmentation.

The more segmented the more secure a network is, but unfortunately also the more difficult to implement and maintain. This is certainly true for traditional virtual local area network (VLAN) built based on traditional bridges/routers/switches/firewalls. With the recent advent of software-defined networks (SDN) and supporting commercial products the task has become easier and making the implementation and management of highly or microsegmented networks much more feasible.

10.4.2 Firewalls

Firewalls protect and control data flows and access to the enterprise network (as a boundary to the outside) as well as between internal network segments. In a sense, firewalls restrict access from one network (segment) to another, or from an enterprise network to the internet, with the goal to improve the security posture of the network.

Firewalls translate and enforce an organization's network security policies via firewall policies and are traditionally implemented as hardware appliances, but can also be virtual or software-based (e.g., as part of a commercial anti-malware product installed on an endpoint). Due to the critical security function of firewalls they always should be hardened so to prevent unauthorized changes that could weaken their security function. Also, firewall change management requires a deliberate planning and implementation process to avoid either extreme (too restrictive or not secure enough) and to assure that the firewall policies are properly maintained over time.

Depending on use case and purpose, firewalls vary widely in architecture and functionality and the various types not only vary in their detection and prevention capability but also in their ability to log events and provide alerts. Obviously, the more complex and capable the firewall, the higher the impact on bandwidth—or, the higher the hardware effort required to maintain a given bandwidth. Some of the more common firewall types are:

- *Packet-filtering firewall*: A fairly simple model that uses basic rules and access control lists (ACLs) to make simple accept/reject decisions on a packet-by-packet level. Although efficient, they require careful design and maintenance of the firewall rules and have limitations with regards to their security capabilities (e.g., they cannot detect an application-specific vulnerability in a data stream that is sent/received as part of the communication between one legitimate application and another). Due to their design and rule-based operation they are very susceptible to vulnerabilities due to configuration error or maintenance drift.

- *Stateful firewalls*: Typically work on the OSI transport layer to provide more advanced and state-dependent filtering (i.e., they are capable of more complex and granular decisions). A common example is that of a user visiting a website. A stateful firewall would allow the traffic from the web site back to the user's browser since it was initiated from there. However, it would reject any random traffic from the outside that was not previously logged within the firewall's state table. Again, there are some security limitations as for example a stateful firewall would not protect from a compromised website visited, knowingly or unknowingly, by an internal user.

- *Proxy firewalls*: Act as a middleman (hence: proxy) and perform a deep inspection of the messages leaving and entering a network. It also isolates internal endpoints from the external network as it replaces the internal address with the address of the proxy (i.e., internal devices are never directly exposed to the external network, they remain anonymous.) Therefore, from a security perspective, proxy firewalls provide two advantages, the deep inspection of messages as well as isolation of the internal network from the external world.

   Actual implementations vary and may provide different levels of inspection. A firewall may be capable of (or configured to) inspect application-level traffic and, for example, make decisions about files attached to an email. But in cases where traffic may come from an unknown application, for example, an FTP request, a firewall may fall back on simpler rules (e.g., based on allowed IP addresses). Although very good for security, proxy firewalls may result in performance (network bandwidth) or functional limitations.

- *Next generation (NG) firewall*: All the previously discussed types have one common limitation, and that is that they depend on known risks and threats or predefined rules (as defined by their policies). They may be able to detect known bad IP addresses, suspicious files, or malware, but will likely fail if confronted with unknown and new threats. Next generation firewalls introduce a new concept by executing programs or opening files in a controlled environment and observe their behavior (commonly referred to as exploding). For example, an email-attached file is captured by the firewall and executed in a virtual environment to see if it behaves maliciously and only if it is benign is it passed on to the email recipient. Obviously, an NG firewall does not solve all security problems, but it addresses the problem of the unknown malicious file. Although, as always in cybersecurity, it is an arms race and cyber adversaries have responded to NG firewalls by delivering malicious content in separate pieces rather than all at once (requiring that the firewall is smart enough to recognize the individual components) or smart malware may detect that it is being executed in a virtual environment and may not detonate its payload.

Typically, firewalls are deployed as boundaries between network segments or between the internal and external network (Internet). However, there are also cases where firewalls are used for just one specific device. For example, any system or device that is protected with a commercial endpoint security solution (antimalware or HIDS/HIPS as discussed in Section 10.1) has a built-in software firewall that should be turned on unless performance or predictability concerns would prevent that. Also, devices that have inherent vulnerabilities that cannot be addressed (e.g., running an end-of-support operating system) can be protected by an external hardware firewall. This would reduce the likelihood of an external attack succeeding, but remain vulnerable to nonnetwork attack vectors, for example USB data carriers (air-gap attack).

From an implementation perspective, firewalls can be added across several security control points and come in a number of form factors (e.g., as a physical device or virtual appliance). Also, many COTS software components, like the OS, anti-alware, or HIDS/HIPS, may come with a built-in software firewall that can be implemented on the device level. Lastly, there are low-cost firewall appliances that can be dedicated to a specific device to add an additional layer of protection (although, from an economic perspective, probably a more feasible approach for imaging equipment than battery operated devices, e.g., infusion pumps). It is critical that the differences and respective limitations between the various types and form factors are understood and be taken into consideration. Simply stating the fact that we have a firewall is not a sufficient statement about security.

10.4.3   IDS/IPS Systems

NIDS and NIPS detect network intrusion based on matching observed behavior against what is defined as unusual or abnormal behavior. IDS is simply focused on detection, logging, and alerting, whereas an IPS would take action and actually interfere with or prevent suspicious activity.

The choice between IDS and IPS depends on the level and type of risk a network operator is willing to accept. An IPS may stop good traffic in a false-positive situation, which may or may not be acceptable. On the other hand, an IDS would insert a human decision into that process so that the preventive action could consciously be decided on, but this would delay action and allow the adversary more time on the network.

Common attack activities that IDS/IPS systems can detect and prevent are external port scans and similar reconnaissance activity, denial of service (DOS) attacks, network policy violations, or activity typically associated with a network exploit.

IPS/IDS solution commonly use traffic rules and/or a set of signatures that would identify abnormal and malicious behavior. Obviously, the quality of detection highly depends on having up-to-date rules and signatures loaded into the system. Modern and more advanced systems introduce an anomaly detection approach that learns from normal network traffic and then establish a threshold after which traffic patterns would be considered abnormal.

Traditionally, IDS/IPS differ from firewalls as the latter relies on a set of static rules (the firewall policies) to manage network connections (i.e., it is a reactive but preventative approach to intrusions). An IDS/IPS only intervenes once an intrusion has been detected. With today's advanced adversaries, either one has its advantages in certain attack scenarios and that is why with today's more advanced systems the lines between these two security technologies is blurring and modern systems contain elements of both.

10.4.4   Security Deception: Honeypots

Honeypots are decoy computers on a company's network that attempt to entice malicious users to attack in order to collect/analyze malicious behavior. They are deliberately set up to be attractive to an attacker, either by showing security vulnerabilities that could be exploited or by pretending to have valuable information stored in them. Commercial offerings are available that help with the setup of honeypots as part of the larger security operation, including the required capture and analysis tools. Some products can even be set up to look more realistic by mirroring existing system types, such as medical devices.

Although honeypots are not necessarily a defensive security tool by themselves, they are a useful tool in the arsenal of an enterprise's larger security strategy and certainly are an important forensic tool. Note that when setting up a honeypot one needs to be conscious of the legal boundaries between enticement and entrapment. I can use a honeypot to track a criminal that is already committing a crime (e.g., broke into my network) but I cannot motivate somebody to commit a crime that they were otherwise not planning on (e.g., by advertising free MP3 downloads).

10.4.5   Data Loss Prevention

DLP solutions are more focused on information protection than security. Commercial DLP solutions provide the capability to identify data that meet certain criteria (e.g., patient data, account numbers, or intellectual property) at rest as well as in motion. Once the data is identified, protective actions can be taken, such as the data being removed/blocked, encrypted, or the user can be alerted.

DLP products provide both protection against malicious activity (e.g., data exfiltration) as well as careless user behavior (e.g., copying data to a USB stick or emailing sensitive data to an outside entity). Products vary in complexity and features and allow an enterprise to implement their information protection rules via DLP policy across traditional endpoints to the cloud as well as transport mechanisms like email or copying data.

In the context of medical devices, DLP products are helpful as they may provide insight on data transfers that are in violation of corporate policy. For example, a medical device manufacturer may copy data from a device for maintenance purposes but may inadvertently copy PHI, identifiers, or network credentials. MDMs may use DLP on their corporate networks to protect their intellectual property (design documents, source code, and research and clinical trial data).

10.4.6   SSL Inspection

One challenge every network security technology faces is that of detecting malicious content in encrypted traffic, be it for security purposes (e.g., IDS/IPS) or privacy protection (e.g., DLP). Although encryption of data is an important security tool, it can also be used by a malicious actor to hide content of data (e.g., malware being sent into an organization or information being exfiltrated). As a simple example, if adversaries don't want to be flagged when exfiltrating source code or patient data, they could just send it via an encrypted SSL or TLS channel. And, unfortunately, that is a technique increasingly used in advanced attacks.

One way to address at least part of the problem is to introduce a network-based SSL inspection system that essentially terminates the encrypted traffic at the network boundary, decrypts and inspects it, and then re-encrypts it before sending it on to the destination. Such commercial tools exist and can be deployed; however, two concerns need to be addressed: degradation of bandwidth and introduction of latency as well as maintaining the level of encryption (i.e., encrypt outgoing traffic at the same strength as incoming).

10.4.7   Security Management Tools

To round off this section, here is a quick rundown of security management tools that may be used on an enterprise network:

- *Compliance and vulnerability assessment*: Scans an IT environment for known vulnerabilities and assesses compliance against IT standards (note, some medical devices may not respond well to such scans, they should be tested and if needed excluded from the scan list before put on the production network);

- *Security risk management tools*: Provides automated assessment of an enterprise's security posture and automates risk management tasks;

- *CMDB*: An IT asset management tool that automates processes around common IT management tasks (deployment, updates, management, etc.) and provides IT asset visibility (again, medical devices may need to be excluded from being managed, especially if such a tool uses agents that are located on the target endpoints);

- *Enterprise security tools (various):* Support common tasks like access control and authentication, automated security training, simulation, and awareness, encryption, and email and web security;

- *Security management, orchestration and response:* These are provided through SIEM systems, SOC, and incident response and remediation.

The above may not be directly applicable to medical devices outside of select cases, but medical device software engineers and service personnel needs to be aware that medical devices will be operated in an IT environment that may use some of these tools. For example, a medical device may be exposed to an asset discovery or vulnerability scan and the device's operation should not be affected by it. Should there be concerns that a device may react negatively to these IT tools, it should be explicitly stated and appropriate configuration and management instructions should be provided.

10.5   Conclusion: How Are MDMs Making Use of These Technologies?

Although the situation is improving, many of the medical devices being sold into the market today still have insufficient security mechanisms. MDMs have often relied on the HDO's security measures, and it was generally thought that network segmentation was enough to secure a medical device. While it is certainly desirable to isolate a networked medical device and firewall it, this doesn't necessarily mean the device is sufficiently secure to prevent:

- Other computers and devices on the network from inadvertently accessing the device and causing it to misbehave;

- Malware introduced by a data carrier (e.g., USB thumb drive) or introduced by an infected laptop;

- A hacker on the hospital network.

A firewall and network separation alone are not adequate security measures. And even these measures may not be available, or adequately implemented, for all medical device deployment and use scenarios (e.g., devices operating in a home health care environment).

Table 10.4 summarizes the applicability of selected key security and which entity typically would implement it (MDM vs. HDO).

Today, MDMs and regulators are embracing the approach of medical devices that are secure by design. However, the long lead-time to introducing updates into already-fielded devices and to incorporating new security technologies into devices that are far along in design and testing means that many more devices are likely to enter the market with insufficient security features. It will take some time to properly secure the fleet of medical devices in the field. As more cryptography solutions become available that focus on addressing the unique requirements of medical device security, MDMs may be able to offload some of these concerns to organizations whose core competency is medical

Table 10.4
Overview of Security Technology Applicability

	Technology	**Use Case**	**Trade Off**	**Note**
MDM	Antimalware	Protect commercial and application software	Resource impact; need for updates; increased attack surface through connectivity; false positives	Suitable for computer-like systems; known entity to customers
	HIDS/HIPS	Protect commercial and application software	Up-front engineering effort to develop policies; limited field flexibility	Suitable to protect commercial OS platforms, even on resource-limited systems
	Cryptography	Protecting confidentiality, integrity, and authenticity	Resource requirements; developer training; protection of keys/ certificates	May require supporting infrastructure to manage keys
HDO	Network segmentation	Separation of critical systems	Effort to manage and maintain; does not prevent physical/local attacks such as USB	May not deter sophisticated hacker; does provide incident containment
	Firewalls, gateways	Separation of critical systems	Does not protect from physical/local attacks, such as USB; requires maintenance	May not deter sophisticated hacker; does provide incident containment
	Anomaly detection	Network-based device security	Relatively new technology but maturing; steep learning curve; risk of false positives	A reasonable alternative to secure devices and IoT networks

device security. This will free up medical device vendors' engineering resources to focus on the functions of the device itself.

Until medical devices are secure by design, and can withstand potentially hostile network and physical environments, inadequate perimeter defenses and reactive security will still be relied upon. Many issues will only be uncovered after they have potentially impacted the function of the device and patients will be at greater risk. While the current state of medical device security remains inadequate, MDMs, the FDA, and companies that provide proactive security solutions are now taking these security concerns seriously and progress is being made. Things are moving in the right direction.

References

[1]   Ferguson, N., B. Schneier, and T. Kohno, *Cryptography Engineering: Design Principles and Practical Applications*, John Wiley & Sons, Hoboken: NJ, Feb. 2, 2011, https://www.schneier.com/books/cryptography_engineering/.

[2]   Moghimi, D., B. Sunar, T. Eisenbarth, and N. Heninger, "TPM-FAIL: TPM Meets Timing and Lattice Attacks," https://tpm.fail/tpmfail.pdf.

[3]   Kelly, S., "Security Implications of Using the Data Encryption Standard (DES)," *The IETF Trust*, 2006, https://tools.ietf.org/html/rfc4772.

[4]   "CWE-338: Use of Cryptographically Weak Pseudo-Random Number Generator (PRNG)," *MITRE*, https://cwe.mitre.org/data/definitions/338.html.

[5]   Genkin, D., A. Shamir, and E. Tromer, "RSA Key Extraction via Low-Bandwidth Acoustic Cryptanalysis," *Cryptology ePrint Archive*, Dec. 18, 2013, https://eprint.iacr.org/2013/857.

[6]   Bernstein, D.,J. Buchmann, and E. Dahmen (Eds.), *Post-Quantum Cryptography,* Berlin: Springer-Verlag, 2009.

[7]   NIST, "Annex A: Approved Security Functions for FIPS PUB 140-2, Security Requirements for Cryptographic Modules," June 10, 2019, https://csrc.nist.gov/csrc/media/publications/fips/140/2/final/documents/fips1402annexa.pdf.

        Bernstein, D., T. Chou, C. Chuengsatiansup, A. Hulsing, T. Lange, R. Niederhagen, van C. Vredendaal, "How to Manipulate Curve Standards: a White Paper for the Black Hat," Jul. 22, 2014, https://bada55.cr.yp.to/bada55-20140722.pdf.

        Bernstein, D., "Curve25519: New Diffie-Hellman Speed Records," Sept. 2, 2006, https://cr.yp.to/ecdh/curve25519-20060209.pdf.

[8]   Cooper, D., S. Santesson, S. Farrell, S. Boeyen, R. Housley, and W. Polk, "Internet X.509 Public Key Infrastructure Certificate and Certificate Revocation List (CRL) Profile," May 2008, https://tools.ietf.org/html/rfc5280.

[9]   Arkko, J., K. Norrman, V. Torvinen, and Ericsson, "Perfect-Forward Secrecy for the Extensible Authentication Protocol Method for Authentication and Key Agreement," Jul. 25, 2019, https://tools.ietf.org/html/draft-ietf-emu-aka-pfs-00.

[10]   Chokhani, S.,W. Ford, R. Sabett, C. Merrill, and S. Wu, "Internet X.509 Public Key Infrastructure Certificate Policy and Certification Practices Framework," Nov. 2003, https://www.ietf.org/rfc/rfc3647.txt.

[11] Moriarty, K., (Ed.), B. Kaliski, J. Jonsson, and A. Rusch, "PKCS #1: RSA Cryptography Specifications Version 2.2," Nov. 2016, https://tools.ietf.org/html/rfc8017.

[12] McGrew, D., H. Igoe, and M. Salter, "Fundamental Elliptic Curve Cryptography Algorithms," Feb. 2011, https://tools.ietf.org/html/rfc6090.

[13] Schaad, J., "Use of the Advanced Encryption Standard (AES) Encryption Algorithm in Cryptographic Message Syntax (CMS)," Jul. 2003, https://tools.ietf.org/html/rfc3565.

[14] Gueron, S., "Intel Advanced Encryption Standard (AES) New Instructions Set," May 2010, https://www.intel.com/content/dam/doc/white-paper/advanced-encryption-standard-new-instructions-set-paper.pdf.

[15] Nir., Y., and A. Langley, "ChaCha20 and Poly1305 for IETF Protocols," May 2015, https://tools.ietf.org/html/rfc7539.

[16] Harris, B., "RSA Key Exchange for the Secure Shell (SSH) Transport Layer Protocol," Mar. 2006, https://tools.ietf.org/html/rfc4432.

[17] Sheffer, Y., G. Zorn, H. Tschofenig, and S. Fluhrer, "An EAP Authentication Method Based on the Encrypted Key Exchange (EKE) Protocol," Feb. 2011, https://tools.ietf.org/html/rfc6124.

[18] Duque, A., R. Stanica, H. Rivano, and A. Desportes, "Decoding Methods in LED-to-Smartphone Bidirectional Communication for the IoT," *Global LiFi Congres,s* 1st Edition, Paris, Feb. 2018, pp. 1–6, https://hal.inria.fr/hal-01683629/document.

[19] Waddell, K., "Your Phone Is Listening—Literally Listening—to Your TV: All in the Name of Serving you More Targeted Ads," *The Atlantic*, Nov. 19, 2015, https://www.theatlantic.com/technology/archive/2015/11/your-phone-is-literally-listening-to-your-tv/416712/.

[20] Perrin, T. (Ed.), and M. Marlinspike, "The Double Ratchet Algorithm," Nov. 20, 2016, https://signal.org/docs/specifications/doubleratchet/.

[21] Grauer, Y., "WikiLeaks Says the CIA Can 'Bypass' Secure Messaging Apps Like Signal. What Does That Mean?" *Slate*, Mar. 8, 2017, https://slate.com/technology/2017/03/wikileaks-says-the-cia-can-bypass-signal-what-does-that-mean.html.

[22] Barker, E., "Recommendation for Key Management," *NIST*, May 2020, https://nvlpubs.nist.gov/nistpubs/SpecialPublications/NIST.SP.800-57pt1r5.pdf

[23] Bluetooth SIG, "Bluetooth Core Specification 5.2," Dec. 31, 2019, https://www.bluetooth.com/specifications/bluetooth-core-specification/.

[24] Afaneh, M., *Intro to Bluetooth Low Energy*, Nov. 22, 2018, https://www.novelbits.io/introduction-to-bluetooth-low-energy-book/.

[25] Afaneh, M. "Bluetooth 5 speed: How to Achieve Maximum Throughput for your BLE Application," Sep. 6, 2017, https://www.novelbits.io/bluetooth-5-speed-maximum-throughput/.

[26] Padgette, J., J. Bahr, M. Batra, et al., "Guide to Bluetooth Security," *NIST*, May 2017, https://nvlpubs.nist.gov/nistpubs/SpecialPublications/NIST.SP.800-121r2.pdf.

[27] Zuo, C., H. Wen, Z. Lin, and Y. Zhang, "Automatic Fingerprinting of Vulnerable BLE IoT Devices withStatic UUIDs from Mobile Apps," *CCS '19*, London, Nov. 11–15, 2019, https://web.cse.ohio-state.edu/~lin.3021/file/CCS19a.pdf.

See also: https://cve.mitre.org/cgi-bin/cvekey.cgi?keyword=bluetooth+low+energy.

[28] Garbelini, M., S. Chattopadhyay , and C. Wang, "Unleashing Mayhem over Bluetooth Low Energy," https://asset-group.github.io/disclosures/sweyntooth/.

[29]   Greenberg, A., "Hacker Lexicon: What Is Perfect Forward Secrecy?" Nov. 28, 2016, https://www.wired.com/2016/11/what-is-perfect-forward-secrecy/.

[30]   https://github.com/greatfet-hardware/quince.

[31]   Walters, P., "The Risks of Using Portable Devices," *US-CERT*, 2012, https://www.us-cert. gov/sites/default/files/publications/RisksOfPortableDevices.pdf.

[32]   Halperin, D., T. Heydt-Benjamin, B. Ransford, S. Clark, B. Defend, W. Morgan, K. Fu, T. Kohno, and W. Maisel, "Pacemakers and Implantable Cardiac Defibrillators: Software Radio Attacks and Zero-Power Defenses," *IEEE Symposium on Security and Privacy*, 2008, https://www.secure-medicine.org/hubfs/public/publications/icd-study.pdf.

[33]   https://www.denx.de/wiki/U-Bootdoc/BasicCommandSet.

[34]   https://github.com/gmbnomis/uboot-mdb-dump.

[35]   O'Flynn, C., "Getting Root on Philips Hue Bridge 2.0," Jul. 12, 2016, https://colinoflynn. com/2016/07/getting-root-on-philips-hue-bridge-2-0/.

[36]   Davidov, M., "Microcontroller Firmware Recovery Using Invasive Analysis," Duo Security, Mar. 26, 2018, https://duo.com/labs/research/microcontroller-firmware-recovery-using-invasive-analysis.

       Riddle, S., "Decapping Chips," Sep. 8, 2016, https://seanriddle.com/decap.html.

       Clark, J., "Don't Try This at Home: Decapping ICs With Boiling Acid," Jun. 24, 2016, https://labs.f-secure.com/archive/dont-try-this-at-home-decapping-ics/.

[37]   Ronen, E., C. O'Flynn, A. Shamir, and A,Weingarten, "IoT Goes Nuclear: Creating a ZigBee Chain Reaction," https://eprint.iacr.org/2016/1047.pdf.

[38]   Witteman, M., "Secure Application Programming in the Presence of Side Channel Attacks," Nov. 2018, https://www.riscure.com/uploads/2018/11/201708_Riscure_ Whitepaper_Side_Channel_Patterns.pdf.

       Wu, M., S. Guo, P. Schaumont, and C. Wang, "Eliminating Timing Side-Channel Leaks Using Program Repair," *ISSTA'18*, Amsterdam, Jul. 21, 2018, https://arxiv.org/ pdf/1806.02444.pdf.

[39]   O'Flynn, C., "A Framework for Embedded Hardware Security Analysis," DalHousie University, Jun. 2017, http://hdl.handle.net/10222/73002.

[40]   O'Flynn, C., and Z. Chen, "Side Channel Power Analysis of an AES-256 Bootloader," *CCECE* 2015 in Halifax: NS, 2015 https://eprint.iacr.org/2014/899.pdf.

[41]   Gamaarachchi, H., and H. Ganegoda, "Power Analysis Based Side Channel Attack," Jan. 2018. https://arxiv.org/pdf/1801.00932.pdf.

[42]   Haislip, H., J. Deltognoarmanasco,, T. Tepp, H. Stockley, D. Pettit, and S. Michels, "Medical Device Isolation Architecture Guide," U.S. Department of Veterans Affairs, Aug. 2009, http://s3.amazonaws.com/rdcms-himss/files/production/public/ FileDownloads/2013-Medical-Device-Isolation-Architecture-Guide-2009.pdf.

11

Select Topics/Deep Dives

This final section will address a range of important topics that have not been covered in the course of the book—mainly topics that relate to the MDM's responsibilities in unique situations that may arise in their relationship with particular customers. This includes MDM support organization responsibilities, MDM incident response activities, specific deployment scenarios in the patient's home or in the case of military health services, and lastly, scenarios where an MDM feels they may not have to share the security responsibility.

11.1   Support Organization Cybersecurity Responsibilities

Previously, we discussed MDM cybersecurity responsibilities within engineering and production processes. Another organizational entity that increasingly has had to take on cybersecurity tasks is the MDM's service and support organization. This entity has a number of responsibilities relating to cybersecurity that are essential to assure patient safety, regulatory compliance, and privacy protections.

It is advisable that MDMs perform a separate and dedicated security risk analysis for their service, support, and maintenance activities, and mitigate risks through appropriate processes, security technologies, and training. These should support meeting the following maintenance and service security objectives: (1) prevent device compromise, (2) maintain the device's security posture over its useful life, (3) protect PHI and other sensitive data, and (4) support detection of and response to security incidents.

The following considerations and general guidelines are not only common security best practices, they are based on practical experience and examples of security incidents that resulted from service activity. For example, the authors are aware of several medical device security incidents with significant impact on care delivery where a USB thumb drive used by a traveling service technician introduced malware to medical devices, resulting in shutdown of medication cabinets or even entire Cath labs. Such a USB attack vector is especially concerning as it can bridge into an air-gapped environment (intentionally or accidentally) or can affect network-segmented devices.

Another example was provided by security researchers who reported that an MDM's download site was compromised by malware, to which any hospital downloading legitimate device software updates was then exposed [1].

To understand the critical nature of support's role in cybersecurity, consider the following specific examples of service activities that impact cybersecurity.

Maintaining device security posture

As part of the MDM's maintenance lifecycle (see Chapter 6), the manufacturer will, from time to time, provide security updates as part of their risk mitigation process (e.g., in response to a newly discovered vulnerability or threat). The HDO may have a formal, contractual relationship with the MDM to provide maintenance service to their devices, which would include cybersecurity updates such as configuration changes or patches to be deployed as a special action (if high priority), or as part of the regular maintenance schedule (if low priority). In case the MDM is not contracted to perform this type of update, service personnel may still be asked to assist customers with planning, managing, and executing what could be quite complex field update projects. In either case, the maintenance activity needs to be coordinated with hospital operations and care delivery priorities and may require complex change management. Change-supporting activities may include planning, supply of update, installation, testing, and training.

Alerting customers about observed cybersecurity risks or actual events and incidents of a medical device

This can be based on a formal relationship and as part of a managed service contract under which the service provider collects and analyzes event logs and alerts. Another possibility is that, during the time of service, a technician may observe indicators (functional compromise or event log data) that a device has been compromised or that a security event is taking place.

Because the manufacturers have visibility across their entire installed base, they are capable of gathering and analyzing event data and observing trends that should be shared with the customer base as part of expected postmarket activities.

Supporting customers in incident-response situations either based on a formal, contractual relationship or on an as-needed basis

These functions include analysis of ongoing incident and support of impact assessment, recovery of device functionality and/or data, final restoration of device to original state, and supporting collection and analysis of forensic data. In the typical case, the HDO will have the main responsibility for incident handling and recovery and typically, the MDM's service organization will provide a supporting role. Incident response activities are discussed in more detail later in this chapter.

Assurance that support activities do not introduce cybersecurity risks or incidents

Local support: service technicians need to be fully trained in cybersecurity best practices and how to maintain a device's (or connected IT infrastructure's) security posture. Service tools they use (e.g., laptops, USB data carriers) need to be free of compromise (e.g., malware).

Remote support: similarly, remote support activities should not introduce any cybersecurity risks to the HDO IT infrastructure. This includes assurance that MDM IT infrastructure is free of compromise; secure and encrypted access (e.g., via VPN); and protection of support user accounts and access (e.g., no default or easy-to-guess user names and passwords; use of multi-factor authentication). Specific guidance for remote support activities is provided through ISO 11633-1 and ISO 11633-2: "Health informatics—Information security management for remote maintenance of medical devices and medical information systems" [2].

Typically, service accounts provide high (if not highest) privilege access and compromised credentials could enable an attacker to have unrestricted access and completely take over a device. Therefore, it is advisable that (a) the design of the device includes a strong and well-planned role-based access control model (RBAC), and (b) that local and remote service access accounts should be equipped with multifactor authentication. Depending on device architecture, this could be based on commercial solutions (e.g., for computer-like devices), or may require a software solution specifically designed for the device that provides that second factor feature (or a comparable security control).

Needless to say, the device should be designed with maintenance and service security needs in mind and provide features that support the above (and similar) use cases. This includes security features (e.g., disable USB autorun), access and authentication (e.g., 2-factor-authentication for high-privilege access), PHI protection during the lifecycle (including EOL data destruction), as well as incident response and recovery/restoration features. Increasingly, HDOs are including postmarket support, incident response, and device restoration in the contractual relationship with the MDM. For more information see Chapter 5.

The review of a service organization's security needs are largely based on the typical scenario of hospital-based equipment. However, many medical devices may not fit that pattern, for example home-based systems (which are further discussed later in this chapter), or patient-worn and implantable devices. These are typically not serviced by dedicated technical personnel, but are often maintained by the patient themselves or by the patient's clinical care provider. Even in these scenarios, the same basic security requirements exist and need to be supported by the device and the device's infrastructure: (1) prevent compromise, (2) maintain security posture, (3) protect sensitive data, and (4) detect and respond to incidents. To enable a nontechnical user to perform these tasks requires special design consideration, as will be discussed in more detail in subsequent sections.

11.2   Incident Response

Incident response (IR) is a critical part of every cybersecurity program. Unfortunately, with today's sophisticated threats, it is not a matter of if a compromise occurs but when. At the time such an event occurs, it is of critical importance for the organization to be prepared in advance to minimize the impacts (on patient safety, data and equipment loss, operational and financial impact, and duration), optimize recovery, execute flawlessly, and enable the capture of valuable forensic evidence. IR is not only a technical task but also includes several important nontechnical steps, including clinical decision-making and internal/external communication.

The most common incident scenario is that in which a fielded medical device at an HDO is subject to a security incident. In this case, the responsibility for managing the incident lies with the HDO. However, there are variations of this scenario, as well as other scenarios, in which an MDM must address or support incident response. These should be covered in an IR Plan:

- IT incident and impact on MDM general IT infrastructure;
- OT incident impacting MDM operational infrastructure (e.g., production); this is related to the previous bullet but generating unique additional concerns(e.g., potential impact on production quality or yield);
- Service and support-related incidents that may also impact customers;
- Customer incident in cases where the MDM is contracted to provide maintenance and support directly to customers, possibly including IT services and cybersecurity management;
- Customer incident managed by the customer where the MDM may be asked for support.

Several of the above are in line with established IR best practices and have been documented in great detail elsewhere, for example:

- ISO/IEC 27035: Information Security Incident Management [3];
- NIST SP 800-61: Computer Security Incident Handling Guide [4];
- NIST SP 800-83: Guide to Malware Incident Prevention and Handling [5];
- NIST SP 800-86: Guide to Integrating Forensic Techniques into Incident Response [6];
- NIST SP 800-150: Guide to Cyber Threat Information Sharing [7];
- NIST SP 800-184: Guide for Cybersecurity Event Recovery [8];
- SANS: Creating and Managing an Incident Response Team [9];
- CMSEI: Handbook for Computer Security Incident Response Teams (CSIRTs) [10];
- ISACA: Incident Management and Response [11].

Therefore, the following discussion will be limited to some general best practice advice as well as aspects that are unique to the medical device ecosystem, as for example provided by the MITRE whitepaper "Medical Device Cybersecurity Incident Preparedness and Response." [12]

The most critical, but also most complex, scenario is when an MDM's device experiences a cybersecurity incident in the field—for example, anomalous traffic on its communication channel is logged and reported to a user or field technician. The MDM must support the affected HDO appropriately, and must document incident, response, and response effectiveness (if applicable). As a reminder, a cybersecurity 'incident' is distinguished from a threat by evidence of an active or potential compromise attempt against a device or the device's network (e.g., a cloud service or connected accessory). Though keeping abreast of incidents, vulnerabilities, and threats are critical aspects of postmarket support, investigating incidents to determine their potential impact is more urgent than investigating newly reported vulnerabilities or threats.

Any security compromise of the device's integrity, availability, or confidentiality is considered an incident—even if the attempt has not (yet) interrupted or degraded a device's essential clinical functions. For example, consider how broadly the EU's medical device regulation defines incidents: "Any malfunction or deterioration in the characteristics or performance of a device made available on the market, including use-error due to ergonomic features, as well as any inadequacy in the information supplied by the manufacturer and any undesirable side-effect." [13] The MDR considers cybersecurity incidents a subset of these. It should be pointed out that in cybersecurity generally, the term incident is viewed even more broadly. For example, NIST defines it as "an occurrence that actually or potentially jeopardizes the confidentiality, integrity, or availability of a system or the information the system processes, stores, or transmits or that constitutes a violation or imminent threat of violation of security policies, security procedures, or acceptable use policies." In cybersecurity, therefore, an incident is considered both an actual as well as a potential occurrence. As with other terms there is, unfortunately, a degree of discrepancy in the use of traditional safety terms as compared to traditional security terms.

Considering the pace of technological change, the ever-evolving threat landscape, and the widening attack surface driven by the demand for increasingly interconnected devices, every MDM should accept the potential for a future cybersecurity incident as likely, if not inevitable. If this has not seemed to be the case in the past, the assumption is that several factors contributed to underreporting, ranging from our ability to actually recognize a medical device cybersecurity incident and correctly attribute it as such; lack of device features that would give us the ability to alert, log, or report compromise attempts; and deliberate underreporting out of reputational and legal concerns.

It is now commonly accepted that the MDM's best practice is to design a device with the assumption that it likely will be involved in cybersecurity incidents during its useful life [14]. Planning for such incidents is the mark of a responsible, mature, forward-thinking organization that has determined to fulfill the security expectations of regulators and, more importantly, of its client HDOs and their patients.

It should be understood that in a typical medical device IR scenario, there are two series of events taking place—the HDO's response to the incident and their triage, recovery, communication, as well as the MDM's response as part of its postmarket responsibilities (Figure 11.1). The former is patient and care delivery-focused, while the latter is design and device lifecycle-focused. Although some postmarket responsibilities are outlined by FDA and other regulators, many of the MDM's postmarket responsibilities will be determined by the contractual relationship between the HDO and the MDM (see Chapter 5). The remainder of this section is focused on the MDM and outlines the major areas that need to be addressed by an MDM to support incident response of a fielded device.

11.2.1   Incident Response Plan

Incident response begins with planning. A complete set of response procedures addressing the various incident scenarios discussed above should be fully approved and in place, and the MDM should be prepared to execute them, before the device is released to market. IR Plans may vary to address, for example, an incident owned and managed by an HDO where the MDM provides only a supporting role, as compared to other scenarios where the MDM would be the

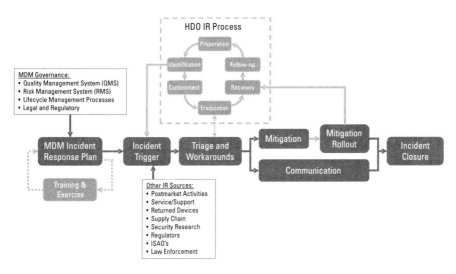

Figure 11.1 MDM IR responsibilities and external touchpoints.

process owner. The MDM's IR plan (or set of plans) should be scalable and flexible enough to address the various possible scenarios.

Incident response planning should harmonize with the MDM's existing risk management system (RMS) and quality management system (QMS). Although they will differ in some respects, procedures for responding to medical device cybersecurity incidents follow a similar basic structure to response planning for IT/infrastructure incidents. According to ISO/IEC 27002, the essential elements of the plan include:

- Procedures governing preparation and response for different incident types and severities. For example, a log report of anomalous comms activity with no apparent impact on the device's clinical function calls for a different type of response and warrants a different level of urgency than a report of a device undergoing manual emergency shutdown due to malfunction or aberrant behavior.

- Procedures for monitoring devices and information sources to detect, analyze, and report incidents.

- Procedures for tracking incident response activities, including activities carried out by the MDM as well as by HDOs and deputized agents in the field (which could include self-reported response activities by end-users).

- Procedures for collecting and handling as well as dealing with forensic evidence.

- Procedures for incident assessment, decision-making, and escalation rubrics.
- Procedures for response, including:
  - Remediation and mitigation determination;
  - Mitigation design;
  - Mitigation deployment and controlled recovery, including deployment tracking and assessment;
  - Response escalation;
  - Communication and reporting, internal and external;
  - Incident closure, including postclosure debrief and follow-up [15].

Depending on the flexibility and maturity of the MDM's organization, these procedures should additionally ensure the competence of assigned personnel—including retraining existing personnel to develop the security capabilities of existing roles—as well as the availability of points-of-contact and reliable, efficient communication channels to all incident-response stakeholders. This requires regular maintenance of contact lists both internal (see Chapter 9) and external (legal and enforcement authorities, regulators, media, resources, and information sharing).

Training procedures for the incident response plan should include mock exercises at least annually. Exercise results should be analyzed, and the results used to inform updates to training procedures and to the plan itself.

11.2.2   Information Sources

MDMs should prepare to receive information pertaining to cybersecurity incidents from multiple channels. Most obvious, of course, is the cybersecurity event logging and reporting designed directly into the medical device system during the secure development lifecycle (Chapter 4). Depending on the event and the device system, however, even this information may reach the MDM in multiple ways. A device system that routinely reports to the MDM's service monitoring infrastructure could inform the MDM directly about potential incidents. In other scenarios, the MDM may be informed directly by their customers (typically the HDO), by device users, field technicians (HDO or contracted), a security incident response team (HDO or contracted), analysis of returned devices, an ISAO (see Chapter 9), or even law enforcement. As information may come in through a variety of channels and may be provided by various sources, the MDM may have to rely on reviewing and correlating reports, which may or may not be clearly reported as cybersecurity incidents.

It will be necessary to train support personnel to recognize keywords (such as security, hacking, attack, infected, etc.) and other indicators in user requests for troubleshooting and support, or complaints about aberrant device behavior, that could point to a cybersecurity incident, and escalate as appropriate. With training, some of these sources—such as HDOs and field technicians—can be channeled through the MDM's existing support system. Others—ISAOs, disclosures by software and hardware suppliers, and in-house security testing—can be captured and assessed as part of the routine activities of the MDM's dedicated product cybersecurity team (Chapter 9).

11.2.3 Investigation

Investigations are primarily an MDM activity initiated by an event trigger. Note that this may shift in the near future, as HDOs add contract language to allow the HDOs to perform penetration testing and other security-related activities.

When incident response is triggered (typically by the HDO as the responsible entity), data gathering activities are initiated to amass as much information about the threat as possible, including an assessment of the quality of the data, and the reputation of the sources of the gathered information. Concurrent with this information-gathering effort, a technical team is enrolled to reviewed gathered data and apply this information in a root-cause analysis (RCA) performed upon representative devices supplied by the MDM. A root-cause analysis is not the same as a vulnerability investigation: the RCA reveals why the vulnerability is present in the fielded system and how to mitigate it. This process can be hampered by insufficient or low-quality threat information, such as which models of medical devices are affected, what version of the software is being exploited, what environmental conditions or configuration are utilized by the threat, and so on.

As the RCA progresses, temporary compensating controls (sometimes called workarounds) should be explored for dissemination to reporting parties, HDOs, ISAOs, and popular media.

Once the vulnerability's root cause is identified, mitigating controls should be implemented in a controlled environment, where the mitigated devices undergo complete revalidation testing to ensure that the mitigation generates no unintentional side effects. The technical team should also use root cause information to explore other similar medical devices created by the manufacturer where this same issue may also be present, but as yet undetected.

It should be noted that all root cause investigations should be performed on nonfielded medical devices, including any fielded medical device that has been returned from the field. No medical devices that have been or may have been attacked should be directly involved in the investigation. Instead, if they become available to the MDM, these devices should be securely quarantined and managed in conformance with a chain of evidence policy.

Communication during all parts of the incident should not be limited to just the team conducting the investigation. At minimum, it is important to let the reporter and other stakeholders know that an investigation is underway. As the investigation progresses, the incident response plan will indicate whether additional parties—such as MDM upper management, ISAO, regulators, or national security agencies—need to be alerted, provide a rubric for making that determination, and should ideally include prewritten templates for communicating the investigation's findings. These templates should have been reviewed, at minimum, by the MDM's public relations team and perhaps by legal as well, as part of the plan's approval process.

The question of what to communicate, when, to whom, and how should never be left up to the individual investigators to decide in the heat of the moment (this applies equally to every stage of incident response).

11.2.4 Triage

In general, the FDA, HDOs, and patients expect that manufacturers will award the highest incident response priority to exploits that may render the device unsafe or compromise its essential clinical performance—or, similarly, which could pivot from the device itself into connected systems and thereby impact patient safety, user safety, or the essential clinical performance of connected systems [16].

Once the root cause has been positively identified, the incident should be triaged based on its scope, potential severity, and exploitability. Confidentiality of protected information categories, such as PI, PII, and PHI should take secondary priority, with exposure of MDM confidential information such as intellectual property taking tertiary priority alongside other threats to the MDM business model (e.g., damage to MDM reputation).

All of this information needs to be presented to the MDM's senior management, along with a proposed plan for addressing the threat in the field. These proposed response types include:

A. Do nothing, identified risk is sufficiently low. (Yes, even when this is the recommended proposal, it is still the incident response team's responsibility to present the facts, findings, and recommendation to senior management!)

B. Once a new version of the software has been validated, update the devices in the field.

C. Perform a voluntary recall, to update and/or replace affected devices.

While a suggested course of action is crafted and suggested by the technical investigation team, the senior leadership will decide on the course of action to be taken.

11.2.5 Roles and Responsibilities

The organization of internal roles and responsibilities, including those that will be involved in managed incident response, are discussed in detail in Chapter 9.

11.2.6 External Communications

Managing communications with stakeholders outside the MDM itself is critical to incident response. Because the general types of incidents that may occur can be anticipated, communications details can and should be preplanned to a great extent. For each incident type, the incident response plan should provide:

1. A high-level summary describing general communications strategy for the incident type;

2. Descriptions of what information should and should not be communicated, to whom, and by what means, for each stage of incident response (investigation, triage and mitigation planning, mitigation rollout, mitigation deployment and efficacy tracking, and closure);

3. A template, including sample language, for each communication.

Communications strategy, specifics, and templates require review and preapproval by at least three groups internal to the MDM: cybersecurity support engineering to ensure technical accuracy and honesty, public relations/marketing/brand managers to ensure messaging is clear, comprehensible to nontechnical audiences, and consistent with the MDM's overall PR strategies, and legal/regulatory to ensure both that communications will reach all parties required by local regulations and that communication contents will limit the MDM's potential exposure as much as possible. Senior management should approve the communications plan only after it has achieved the recommendation of these three groups. Traditionally, many MDMs leaned towards communicating less rather than more. However, from a business impact perspective it is now more commonly accepted that an open communication strategy is a sign of an MDM's cybersecurity maturity and responsible practices.

Depending on the impact and scope of the incident, specifics of the response may vary. But generally, during incident response, external communication generally takes place between the MDM and six other stakeholder categories. These include those who are or who may at some point be directly affected by the incident—HDOs and patients—as well as those who are or who may become indirectly affected—MDM shareholders/owners; regulators and government safety and law-enforcement agencies; industry counterparts with systems, devices, and infrastructure which could be affected by means of the same exploit (or the same underlying vulnerability) through ISAO participation; and the public at large, who are communicated with via the press and any means the MDM uses for direct publication, such as on its corporate website, social media, or circulated news bulletin. Each of these groups has a particular interest in the incident and in the MDM's response which differs greatly from all of the others, as well as differences in the incident-specific information that is appropriate, advisable, and perhaps legal for the MDM to provide them. Consequently, not only what is said, but how and to whom, must be evaluated before any incident actually occurs for the MDM to maintain and enhance its business position during and following incident response.

Yes, it is possible for a well-prepared MDM to actually enhance its business position due to exemplary response to a cybersecurity incident! Remember, we are no longer operating in a legal and business environment in which MDMs are punished or denigrated for having a device on the market that falls prey to a vulnerability exploit. If the MDM developed the device responsibly (i.e., securely), and if it responds appropriately to the incident, it will immediately become an example for consumers and professionals alike as an object lesson in how to do it right!

Effective communication with those who may be directly affected by the incident is also critical for minimizing unsafe unilateral actions that HDOs, end users, or other field personnel might consider taking in the absence of timely information concerning the incident and the MDM's response plan. For example, an HDO might falsely believe that it needed to protect other critical infrastructure by transferring a patient off of a compromised device it believes could infect other systems, which could cause harm to that patient, when all the while the MDM could have a less-risky mitigation option that it had simply failed to communicate clearly or in a timely manner to the HDO. The MDM should tightly coordinate and communicate current activities surrounding the investigation and work with the HDO to develop alternatives and compensating controls until the root cause and mitigating response are ultimately determined.

To that end, an MDM should consider well in advance how it intends to communicate with certain groups. Communications to regulators and to the MDM's ISAO will generally follow established norms and expectations. Communications to HDOs, patients, and other directly affected persons will largely be determined by the specifics of the incident itself, the mitigation plan (see below), and the messaging goals established by PR. However, advance thought should be given to communications to MDM owners and shareholders, to ensure that they are accurately and appropriately apprised without downplaying the incident, oversharing technical details, or needlessly exposing any party to legal liabilities.

Communications to the press also require consideration well before any event occurs. Communication content should be prepared for such events over the course of several months and with rigorous review cycles being performed across a team of multiple disciplines, including technical, customer-facing communications, legal, regulatory, sales, and support.

It is recommended that the MDM's marketing, PR, or media relations personnel cultivate working relationships with reporters who cover cybersecurity and medical devices in key publications well in advance of any incidents, with the goal of establishing channels through which incident-related information can be released at the appropriate time. This will better position the MDM to influence the public narrative and greatly reduce the chances of the incident becoming sensationalized or reported inaccurately.

11.2.7   Mitigation Plan

The steps required to design and execute a successful mitigation plan for software-based incidents are described in detail in Chapter 6. Incidents stemming from threats to hardware components may, in some cases, be able to be mitigated in the field by temporarily disabling vulnerable hardware features (by unplugging a comms cable, disabling a mobile device or wireless channel, etc.) until a long-term solution can be delivered. If the incident is isolated to a single device or device network, it may be possible to replace the compromised device(s) with clean or preupdated units, as appropriate for the incident type, while the compromised device is investigated under quarantine, perhaps avoiding the need to initiate a full recall.

The mitigation plan should consist of two elements:

1. A set of preapproved generic templates describing the general procedure by which various kinds of incidents shall be mitigated, which will be used on an incident-by-incident basis.

2. The specific actions that will be performed, as well as the tools, procedures, and order that will be used to perform them, to mitigate the specific incident. These elements should be drafted by security engineering in conjunction with engineering proper, and reviewed and approved by the Chief Security Officer (or equivalent), as well as other key stakeholders as described in Chapter 9, before it is enacted.

Since time is essential for successful incident response, having a mitigation plan as developed and as detailed as is practical in advance of product release is strongly recommended. Minimizing the quantity and depth of decisions that must be made under the pressure of a live incident or exploit-in-progress is the best way to avoid mistakes or omissions in the mitigation.

Both (1) and (2) should be risk-assessed for their potential impact on the device's safety and essential clinical functions. The MDM's existing risk management rubrics should work for this purpose, enabling the MDM to weigh the risks of proposed mitigations against one another as well as against the current state of the device.

Because of the incident, the device under investigation can no longer be considered to be at the baseline risk level established during premarket approvals. Depending on the potential scope of the incident—does the exploit affect a single device, a single network, an entire product line?—cybersecurity artifacts in the DHF will need to be updated based on refreshed pre- and postmitigation scoring as part of incident closure (further discussed below).

Note that for a large scale and complex incident, the investigation and response may involve multiple organizations, for example, several HDOs or even MDMs, regulatory and law enforcement could be affected (e.g., during the WannaCry ransomware attack), and that a complete response may even cross industries and national borders.

The mitigation plan should also define the emergency actions that may be taken or recommended by support personnel who become aware of an imminent threat or exploit-in-progress. Preauthorized emergency actions should be limited in scope and risk-assessed just like the rest of the mitigation plan.

The goal of the emergency actions section of the mitigation plan is twofold. It exists to authorize swift action to address the incident as early as possible and so, hopefully, to prevent its severity from escalating or its impact from spreading. And it exists to limit the activities of those who might take such actions to controlled activities that can be evaluated in advance of incident discovery.

Finally, the mitigation plan or applicable elements should be communicated to those it will impact—both during training of the internal roles described in Chapter 9, and during the incident itself, to those involved. In terms of outward-facing communication, it is important that affected users and other field personnel be informed as early as possible that the MDM has a plan, first of all, along with any emergency action being taken; and then any plan specifics determined to be appropriate or useful to external parties once those specifics have been approved internally.

Amid the pressure of incident response, it can be easy to overlook or undervalue the important role of nontechnical communication to nontechnical stakeholders, if only for the purpose of providing reassurance in the moment and staying well ahead of the public news cycle, which could become sensationalized. The MDM should always bear in mind its goal is not merely to address the incident on a technical level, but to enhance its market position through every step of response.

11.2.8 Mitigation Rollout and Tracking

Deployment of software/firmware mitigations is described in detail in Chapter 6. In general, and similar to any other software/firmware update, the phases of rollout are:

1. Preparing the update apparatus, including the management server and update package;

2. Preparing and confirming the prepared status of target devices/device components to receive the update (e.g., confirming that taking the device's essential clinical functions offline for the time required to deliver and install the update will not put users at risk);

3. Confirming the update installed and runs successfully or, if it failed, that the device reverted to its previous state;

4a. If the update failed, reinitiating the update procedure or cancelling the update and authorizing the device for return to normal use, if deemed safe to do so;

4b. If the update succeeded, confirming its success and authorizing the device for return to normal use, if deemed safe to do so.

The activity level required from users or field personnel will vary based on the device and the mitigation. For incidents that do not feel urgent or high-priority to users, but which require user activity to mitigate, the MDM may find it helpful to establish incentives for complying with the update process. This is generally less expensive, more efficient, and may be better received by users and the public versus sending its own agents to perform the update activities at each device location (which in certain cases may be impossible anyway). A possible exception may be an on-site mitigation activity which is too technically complex for the average user to perform, or which could expose sensitive information to an inappropriate party.

Tracking the rollout progress includes status reporting, both from the field to the MDM (either by the device itself, as part of the update procedure, or by users or field personnel), within the MDM's own organization, and by the MDM to external stakeholders, including the ISAO, regulators, the media, and so on. As with every stage of incident response, communication is key!

11.2.9   Coordinating with the ISAO during Incident Response

It must be assumed, until proven otherwise, that an incident that affects one device at one customer site also indicates a risk to other customers. Therefore, responsible and timely disclosure of incident-related information is of the essence to enable other participants to be proactive in preventing a similar occurrence. Moreover, communicating a security event to all affected HDOs can be logistically difficult; an ISAO facilitates this communication.

Both MDMs and HDOs should be members of and can use ISAOs to help with this process. The FDA encourages MDMs to actively participate in ISAOs and gives the MDM greater autonomy in the communication process (among other incentives) for doing so [17].

ISAOs have been further discussed in previous chapters. The purpose and organization of an ISAO is described by CISA [18], and a list of approved ISAOs is provided via HSCC [19] and the ISAO Standards Organization [20].

11.2.10 Incident Closure

The incident response plan should include criteria for determining when the MDM has brought an incident to a close, a procedure for officially declaring the incident closed, and a procedure for initiating follow-up/debrief. Depending on the type of incident and its impact, the MDM may need to comply with regulatory processes like device recall or field replacement action.

Incident closure criteria should include a rubric that defines whether the MDM's response to the incident was appropriate for the incident type, was warranted by its investigation, and was communicated appropriately to stakeholders. It should further include a threat matrix, similar to the design vulnerability assessment, which evaluates the pre- and postmitigation risk status of the device for any incidents for which mitigating action was deemed necessary.

The incident response plan should include a template for a final incident report, which describes succinctly, but with sufficient technical detail, (a) what the incident consisted of, (b) how the MDM became aware of the incident, (c) what steps the MDM took to investigate the incident, (d) what the investigation determined, (e) what mitigation the MDM devised to address the incident, if applicable, (f) a timetable charting successful mitigation deployment over time (e.g., within x period of time, y percentage of affected units had successfully applied the mitigation, growing to z percentage of affected units after 2x time, and so on, up to the date of the report), (g) any follow-up steps, if applicable, including postincident follow-up (described below), and (h) the MDM's justification for declaring the incident closed (based on its established closure criteria). Each section should include references to detailed documentation for each response activity. The report should be prepared by the incident response team, approved, then issued to all appropriate stakeholders, including senior management, the ISAO, and regulators. A press release based on the final incident report can be prepared and released to the MDM's developed media contacts (described under external communications, above), coincident with the MDM's final incident communication to patients.

In order to populate section (g) of the final incident report, the MDM will need to identify its post-incident follow-up activities. The general trend of recent and forthcoming cybersecurity standards and regulations calls for MDMs to adopt a posture of continual improvement regarding cybersecurity management, just as they already do with quality management. Incident response follow-up is a major area for such activities. Depending on the incident type and investigative findings, the MDM may deem it necessary to trigger a

CAPA-like process so that findings can be fed back into the MDM's design process or organizational practices, quality system processes, manufacturing controls, and the incident response plan itself. As the threat landscape evolves, so must the MDM's cybersecurity activities. Responsible postincident follow-up provides excellent opportunity for the MDM to exercise, and to demonstrate, its growing maturity.

11.3 Unique Use Environments

11.3.1 Home Care

Security in the home setting has its own set of challenges. Typically, the home care or telemedicine device is placed in an environment with a lower security posture (home network) and transmits/receives data via channels that are neither under the manufacturer's nor health care provider's (or medical service provider's) control (i.e., home and public networks).

Based on their function, most home-based devices are less of a patient safety concern, but there would certainly be privacy concerns with any device. Also, as technology progresses and pressure to reduce health care costs continues, we will see more critical-care devices in patient homes, for example, dialysis systems or monitoring of vital signs.

Common infrastructure for home-based devices include a central hub (gateway) that acts as a data aggregator across all devices, allows management of devices and data flow, and can be used to collect additional patient information (e.g., to confirm identity).

A hub can be a custom system, which allows security and privacy features to be designed into it and therefore enforces and assures a certain security baseline. However, the hub function may also be provided by an app on a smart device, which means that security and privacy features would need to be implemented via the app or even by the device itself.

Typical security needs requiring support and examples of how they could be realized include:

- Privacy and confidentiality (e.g., encryption);
- Security and protection (e.g., firewall);
- Data flow management (e.g., via configuration interface);
- Data association to device and/or patient (e.g., via configuration interface and message signing);
- Maintenance, updates, and patches (e.g., as push function or via user interface);
- Configuration management (e.g., via user interface or remote access);
- Access and authentication (e.g., role-based access control, biometrics).

None of these requirements are necessarily unique to home-based devices, but their location, network connection, and operator capability make their implementation challenging. Ideally, the device as well as the supporting hub should be designed to be sufficiently secure while assuring privacy, usability (for patients and caregivers), availability, remote control, configuration management, and support for the desired connectivity.

Regulatory requirements affect the degree to which such information needs to be protected, which in turn determines the security features and controls required. This may vary based on region (e.g., GDPR for Europe, HIPAA for the United States, or even specific state laws as, for example, California's), as well as type of data (e.g., fitness data, general health data, or data related to specific health scenarios).

Evolving new technologies will continue to change the picture of home care. IoT-based technologies will enable new devices and solve problems that previously could not be addressed. The advent of 5G public networks will make it easier to implement some of the connectivity requirements, but it will also offer new opportunities to adversaries for attacking devices directly.

Consumerization creates another challenge, in that the introduction of consumer devices not only creates new data points into a patient's profile (e.g., cardiac data collected by a smart watch), but also new regulatory and privacy challenges. A patient's exercise history may be relevant to a clinical consult, but it may come from a less controlled device or app. In particular, nonmedical consumer fitness devices have raised significant privacy concerns in the past [21].

A more extensive discussion of the spectrum of personal health devices (PHD), their security requirements, architectural models, and standards mapping has been provided in an IEEE whitepaper [22].

11.3.2 Military Health Services

Another unique environment is that of military health care services; for example, the U.S. Department of Defense (DoD) health services or the Department of Veterans Affairs (VA). Besides being responsible for providing health services to inactive and active military personnel (including combat missions), they typically also support senior political figures including heads of state. Although many of the previously discussed cybersecurity fundamentals apply equally to this environment, it is unique in its size and complexity as well as the unique broad and complex attack surface it offers, making it an attractive target to adversaries with military, political, or criminal motivation. As a result, their cybersecurity requirements tend to be more stringent than that of other HDOs.

Military health services also pose some unique logistics requirements on MDMs. One example is the need to maintain medical devices that are held in storage in support of future combat readiness or disaster response missions.

This means that MDMs may need to provide warehouse-level support and potentially maintain devices in storage, including software updates.

Due to their significant purchasing volume, both VA and DoD hospitals are significant customers to MDMs and therefore, their purchasing requirements tend to drive MDM design decisions. Both were early in recognizing the cyber risks of medical device infrastructure and, not surprisingly, both the U.S. DoD and VA hospitals were early advocates on the topic of medical device cybersecurity and have developed procurement and implementation requirements (and continue to do so). Specific examples include:

- Department of Veterans Affairs VA DIRECTIVE 6550: Pre-Procurement Assessment for Medical Device/Systems, February 20, 2015;
- Department of Veterans Affairs: Medical Device Isolation Architecture Guide, Version 2.0, August 2009;
- Department of Veterans Affairs: Design Patterns Privacy and Security. Medical Device Security, Version 1.0, January 2017;
- Defense Information Systems Agency DISA: Security Technical Implementation Guides (STIGs);
- Department of Defense: DOD 5220.22-M, National Industrial Security Program Operating Manual, February 28, 2006;
- Department of Defense: Medical Devices Security Technical Implementation Guide. Version 1, Release 1, July 2010;
- Department of Defense: DOD Directive 7730.65, Department of Defense Readiness Reporting System (DRRS), May 11, 2015.

In addition to military-specific purchasing requirements, by extension other government IT security standards may need to be met, for example:

- FIPS PUB 200, Minimum Security Requirements for Federal Information and Information Systems;
- FIPS PUB 199, Standards for Security Categorization of Federal Information and Information Systems;
- NIST SP 800-18, Guide for Developing Security Plans for Federal Information Systems;
- NIST SP 800-37, Guide for Applying the Risk Management Framework to Federal Information Systems.

An MDM may need to demonstrate compliance with the above as required as part of the purchasing process, including demonstrating compliance of internal processes and documentation as well as specific certifications to meet standards.

11.4   Common Cybersecurity Excuses and Myths

After many decades of combined experience working with and for MDMs as engineers and security specialists, the authors have experienced just about every excuse for why the MDM didn't secure their medical device. Ironically, it sometimes seems like more time and money were spent on excuses than it would have taken to actually secure the device!

Although we have seen improvement, the capacity of some management (from midlevel to senior) for willfully ignoring security has often been nothing short of astonishing. The level of self-deception entailed sometimes rivals the proverbial ostrich's head in the sand approach to threats. (Ostriches do not in fact put their head in the sand in response to threats. Any animal that behaved in this manner would already be extinct, which in and of itself should serve as a lesson to business leaders.) This is why it is such a pleasure to work with MDMs who address security as an essential aspect of their products and business, and actually help to improve the overall industry. This is, in the end, an issue of executive leadership, the right business priority, and an organization's cybersecurity culture. We need more like these leaders!

We have also seen good people at some MDMs reach a point of such frustration with their organizations that they leave or give up trying to improve the organization's security posture. This section of the book is for those people: wherever you are, at whatever organization you work for, take comfort in knowing that your situation is not unique. Ten years ago, you would have been isolated and fighting a lonesome battle, but now you are in good company with most of the leaders of the medical device industry and the worldwide regulatory agencies working to improve the security of medical devices. So—fight the good fight! This section is dedicated to you and your efforts.

Years ago, management ostriches could be explained away by the general lack of market demand, as well as the lack of security awareness and training that existed for engineers and for the MDM community in general. However, that has not been the case for several years, and it is becoming obvious that this was never about ignorance, but always about not wanting to fund the extra effort necessary to secure medical devices. This does not speak well for where our industry has come from. As we see many members of the MDM community fully embracing security in medical devices, it makes the contrast with those that reject it even starker.

Excuses we have seen (and, sadly, still continue to see on a regular basis) include:

"We do not need to secure our medical device!"

This speaks to such a vast level of incompetence that it is always disheartening to us when we hear it. What the speaker is saying is that he or she has ignored all United States and international regulations about medical device security for years, they are ignoring widely reported security incidents relating to medical devices in the news, and they don't really care if their company's business is damaged or if their patients are harmed by their medical devices as long as short-term profitability is achieved. It is very difficult to make headway with someone like this, as they will not acknowledge moral concern for the consequences their actions may create.

This is a good place to point out that if you don't have the cooperation of at least some members of the senior management team, your efforts are guaranteed to fail. If you can't get the management ostrich in question overruled by someone in a position of greater authority, the best approach is to leave the employment of the company or drop them as a client. That manager has hired you as a token nod toward security and will not hesitate to paint any negative security events that later befall as your fault if he or she finds it expedient to do so.

"Our engineers only write secure code!"

We love this one. It always comes from a pompous senior executive who doesn't really understand security (or engineering). The hardest part is trying to keep from laughing when we hear it! (We're not always successful!) What this person is really saying is some variant of, "I'm a senior executive and I know more about this than you do and my department is handling it. Even though we have no processes in place to track or measure it, trust us, we're doing a perfect job." It also displays the ignorance that all vulnerabilities are software-based.

Once you start tunneling into specifics, these managers start getting quite evasive and push back about not knowing all the details—but they are confident that their engineers are doing this thing called security, even though they cannot point to what processes or artifacts are necessary to design, develop, and manufacture a secure medical device system. And when we then suggest that we get someone from their engineering group to ask, just to clarify their activities, surprisingly, this person or group cannot be reached for questions!

Much like the executive who believes they don't need to secure their medical device, this company is very difficult to make headway with, and typically has a lot more issues than just security. If you ever get to see their source code, you will almost certainly be shocked at its lack of quality.

"Once the source code is compiled, nobody can read it!"

This is one of those odd myths that seem to permeate the software developer's industry. How otherwise skilled software engineers can proclaim such utter hogwash is beyond us. All languages have cheap, readily available disassemblers and decompilers that can be used to quickly and easily expose the intellectual property, algorithms, and security mitigations and methods in the software. Just compiling your code (native or bytecode) in no way protects it from exposure or corruption.

"Nobody can hack our proprietary [fill-in-the-blank]!"

We hear this statement on something close to a monthly basis, stated as a fact by an MDM that their proprietary interface is secure, simply because it is proprietary. And, some engineers may even equate proprietary with encryption, a myth we have hopefully dispelled. Security professionals reverse engineer proprietary technology all the time. Your prized communications protocol really isn't that unique. Yes, there have been examples of security researchers who were stopped dead in their attack of a protocol by such simple technical details as data whitening [23], where long sequences of 1s will have a 0 bit inserted into the long sequence, likewise, long sequences of 0s will have a 1 bit inserted into the long sequence. This can be confusing to nonengineers who are just sniffing the RF signal and trying to make sense of it. However, anyone with any RF technical knowledge will immediately filter out such whitening to restore the original bitstream.

With the introduction of cheap SDRs (some as low as $20), and a whole host of free programs [24] to control the SDR, the process of reverse engineering [25] an RF protocol can be measured in minutes [26]. We can reverse engineer any frequency (from kilohertz to gigahertz), any modulation scheme, and increase the range of any low power RF to the point where it could be weaponized to attack patients in a brush-by attack, where the target device is only in range for a couple of seconds. None of these supposed mitigating controls are actually going to stop a determined, competent attacker. This is equally true for the wired variety of communications as well. Sniffers exist for every type of electrical connection you can find. Even less mainstream communication mediums, such as inductive communications that have been typically utilized in certain implantable medical devices, while being wildly diverse and unstandardized, can be reverse-engineered with the same tools used for RF reverse engineering, plus a few other additional tools.

All proprietary interfaces share a common trait: there is absolutely no real cryptographic security being utilized—not authentication, not integrity checking, not encryption.

One final point worth noting is that if you submit any documentation to the FDA that claims you are secure because of your proprietary interface, we guarantee you will be questioned in detail and ultimately rejected.

"Our data is anonymized."

Don't be afraid of encryption.

- It is not difficult to perform;
- There are many good cryptographic libraries or software tools available;
- If you're using hardware-accelerated cryptography, it doesn't take long to perform;
- Even pure software solutions are going to be measured in milliseconds (depending upon your hardware);
- There are good external components such as the Microchip ATEC-C608A that can perform the operations in hardware if your CPU doesn't support it.

Really, the hardest part of encryption is the key exchange and storage. So why, then, will so many MDMs argue endlessly that the data they are collecting about a patient is not PHI, instead of just encrypting the data? This is a rhetorical question: we have no sensible answer. The argument usually centers around the MDM not including a Social Security Number with the data, an address or a phone number, and so on. They attempt to make the argument that they have deidentified the data via the HIPAA Breach Notification Safe Harbor rule (see Chapter 4 for more information on Safe Harbor) [27]. Fortunately, with the increase in HDO contracts that call for encryption of all data, this argument will not survive for much longer.

The takeaway for this section is that you simply need to encrypt the data, PHI or not. It really isn't that difficult.

"The advantages of our device outweigh the insecurities of the device."

Do they really? Considering that security mitigations exist to protect your device's essential functionality (integrity and availability) and to protect patient and business-critical information (confidentiality), what advantages can your device possibly provide once it has been compromised? Moreover, wouldn't your competitor's secured medical device also deliver similar advantages to the market without the risks being posed to patients by your unsecured device?

"We didn't have to secure the last device we made."

This is actually a valid point. Given the pace of recent changes in security regulations, it is possible that your company didn't have to secure its previous medical device to get premarket approval. However, knowing what you now know about what is going on in the world with cybersecurity and associated attacks, you have to ask yourself,

- "Do I feel comfortable knowing that my insecure device is out in the market? Would I want a loved one to use one of these devices, or would I want to use it myself?"
- "Is there intellectual property on that device that could be extracted by a competitor?"
- "Is there a consumable product utilized as a part of that device that a competitor could clone?"

Yes, the regulatory environment has changed, and it is good for your business and the health and safety of your patients.

"The FDA just doesn't understand our particular situation/device."

A less common form of the hubris of the aforementioned senior executive-type can be found in companies that just won't accept that the FDA has the right to regulate their medical device, and thus enforce patient safety over the company's profitability and/or continued existence. While it is less common, it isn't as rare as most of us would suppose.

This manager believes that they (or the company's lawyers) will successfully argue for a special exemption from the law due to the uniqueness of their device or company. What he or she doesn't understand is that (A) this simply isn't true, and (B) security activities are now part of the refuse to accept list, or prescreening criteria, for all premarket submission applications to the FDA. What this means is that the company that refuses to take cybersecurity requirements seriously will never get to make their case at all.

"Our legal risk is lower if we don't take responsibility for security."

This argument has been presented to one of the authors by an MDM's legal department. They laid out their argument that if they implement security processes and technology, they would be taking on responsibility for security. Therefore, even if it would be much less likely that a security incident would occur, the MDM would now be legally liable since they are responsible for the security features. Hence, the MDM would be better off not to implement security to reduce their legal liability.

What this twisted thinking fails to take into account is existing regulations, which in most markets worldwide declare that the MDM already has the responsibility to secure their device. Lack of security in no way limits or avoids that responsibility; it simply ensures that the MDM will be denied regulatory approval and market entry.

"We agree with what you are telling us about cybersecurity, but we still don't see the business case."

Again, a response by an MDM to one of the authors, stubbornly continuing in denial of the regulatory environment, threat landscape, and market opportunity we've outlined in this book. The denial may be baseless, but in some places, it persists in the face of all reason. However, we do hope that growing customer demand for security and increasing regulatory and reputational risks will eventually make this argument go away.

11.5   In Closing

Our hope is that you can use some of these examples to help improve the product security posture at your organization, but if not, we hope they at least cheered you up! Although we're ending this chapter and book on a lighthearted note, these examples show that we are still far from cybersecurity for medical devices being generally accepted as a business priority and a must-have deliverable that requires support across the entire value chain. Only a mature, well thought through, and carefully implemented security lifecycle process can accomplish this.

We, the authors, as well as our contributors, have spent many years of our careers on this topic and have seen regulators, device manufacturers, security researchers, and healthcare delivery organizations take significant steps forward—but as you were always told as a kid on a long car ride, "we are not there yet." Not by a long way.

We do not intend to contribute to the headline-driven "the sky is falling" view of the problem. Even though some news outlets, and even some security researchers, try to convey that we are in a world-ending crisis, we don't believe that we should panic. But clearly, the scope of the problem and risk to patient safety and national health systems dictate that we do need to proceed with a sense of urgency and take a systematic and deliberate approach to address this challenge, as outlined in this book.

We have not yet seen direct evidence of patient harm that can be definitively attributed to a cyber incident of a medical device (admittedly, we also have not answered the question "how can you tell?"). However, we have to assume that once such an event occurs, the public discussion will change and the impact on business and patient care will be urgent and significant. As a collective industry we need to make sure that we have made as much progress as possible by the time this happens.

As we finish this book (in mid-2020), the world faces a health crisis such as we have not seen in a long time. The resulting changes to our healthcare system will accelerate the digitalization of care delivery and will require an even stronger focus on cybersecurity. We hope that this book will help clarify the right course and further our progress on the path to more secure devices.

References

[1]   Fu, K., "Click Here to Download Your AVEA Ventilator Software Update. Trust Me," Archimedes Center for Medical Device Security, Jun. 8, 2012, https://secure-medicine. blogspot.com/2012/06/click-here-to-download-your-avea.html.

[2]   International Standards Organization, "ISO/TS 11633-1:2019 Health informatics—Information Security Management for Remote Maintenance of Medical Devices and Medical Information Systems—Part 1: Requirements and Risk Analysis," Aug. 2019, https:// www.iso.org/standard/69336.html.

      "ISO/CD TR 11633-2 Health informatics—Information Security Management for Remote Maintenance of Medical Devices and Medical Information Systems—Part 2: Implementation of an Information Security Management System (ISMS)," https://www.iso.org/ standard/78861.html.

[3]   International Standards Organization, "ISO/IEC 27035-1:2016 Information technology—Security Techniques—Information Security Incident Management—Part 1: Principles of Incident Management," Nov. 2016, https://www.iso.org/standard/60803.html.

      ISO/IEC 27035-2:2016 Information Technology—Security Techniques—Information Security Incident Management—Part 2: Guidelines to Plan and Prepare for Incident Response," https://www.iso.org/standard/62071.html.

[4]   Cichonski, P., T. Millar, T. Grance, and K. Scarfone, "SP 800-61 Rev. 2 Computer Security Incident Handling Guide," *NIST Computer Security Resource Center*, Aug. 2012, https://csrc.nist.gov/publications/detail/sp/800-61/rev-2/final.

[5]   Souppaya, M., and K. Scarfone, "SP 800-83 Rev. 1 Guide to Malware Incident Prevention and Handling for Desktops and Laptops," *NIST Computer Security Resource Center*, Jul. 2013, https://csrc.nist.gov/publications/detail/sp/800-83/rev-1/final.

[6]   Kent, K., S. Chevalier, T. Grance, and H. Dang, "SP 800-86 Guide to Integrating Forensic Techniques into Incident Response," *NIST Computer Security Resource Center*, Aug. 2006, https://csrc.nist.gov/publications/detail/sp/800-86/final.

[7]   Johnson, C., M. Badger, D. Waltermire, J. Snyder, and C. Skorupka, "SP 800-150 Guide to Cyber Threat Information Sharing," *NIST Computer Security Resource Center*, Oct. 2016, https://csrc.nist.gov/publications/detail/sp/800-150/final.

[8]   Bartock, M., J. Cichonski, M. Souppaya, M. Smith, G. Witte, and K. Scarfone, "SP 800-184 Guide for Cybersecurity Event Recovery," *NIST Computer Security Resource Center*, Dec. 2016, https://csrc.nist.gov/publications/detail/sp/800-184/final.

[9]   Proffitt, T., "Creating and Managing an Incident Response Team for a Large Company," *SANS*, Jul. 18, 2007, https://www.sans.org/reading-room/whitepapers/incident/ paper/1821.

[10] West-Brown, M., D. Stikvoort, K.P. Kossakowski, G. Killcrece,R. Ruefle, and M. Zajicek, "Handbook for Computer Security Incident Response Teams (CSIRTs)," *Carnegie Mellon Software Engineering Institute*, Pittsburg, PA: April 2003, https://resources.sei.cmu.edu/ asset_files/Handbook/2003_002_001_14102.pdf.

[11] ISACA, "Incident Management and Response," 2012, https://www.isaca.org/bookstore/ bookstore-wht_papers-digital/whpimr.

[12] Connolly, J., S. Christey, R. Daldos, M. Zuk, and M. Chase, "Medical Device Cybersecurity Regional Incident Preparedness and Response Playbook," *MITRE Technical Papers*, Oct. 2018, https://www.mitre.org/publications/technical-papers/medical-device-cybersecurity-regional-incident-preparedness-and.

[13] EU Medical Device Coordination Group, *MDCG 2019-16: Guidance on Cybersecurity for Medical Devices*, Dec. 2019, p. 29.

[14] U.S. Food and Drug Administration, "Postmarket Management of Cybersecurity in Medical Devices," *FDA Guidances*, Dec. 28, 2016., p. 13.

[15] International Standards Organization, "ISO/IEC 27002:2013 Information Technology—Security Techniques—Code of Practice for Information Security Controls," Oct. 2013, https://www.iso.org/standard/54533.html.

[16] FDA, "Postmarket Management...", p. 15.

[17] FDA, "Postmarket Management...", pp. 25–26.

[18] https://www.cisa.gov/information-sharing-and-analysis-organizations-isaos.

[19] https://healthsectorcouncil.org/hic-miso/.

[20] https://www.isao.org/information-sharing-groups/.

[21] Hilts A., C. Parsons, and J. Knockel, "Every Step You Fake: A Comparative Analysis of Fitness Tracker Privacy and Security," *Open Effect Reports*, 2016, https://openeffect.ca/reports/Every_Step_You_Fake.pdf.

[22] Fischer, C., and N. Hamming, "IEEE PHD Cybersecurity Standards Roadmap," *IEEE*, Apr. 30, 2019, https://ieeexplore.ieee.org/document/8703258.

[23] Freescale Semiconductor, "Implementing Data Whitening and CRC Verification in Software in Kinetis KW01 Microcontrollers," Jul. 2015, https://www.nxp.com/docs/en/application-note/AN5070.pdf.

[24] Benchoff, B., "Shmoocon 2017: A Simple Tool For Reverse Engineering RF," *Hackaday*, Jan. 15, 2017, https://hackaday.com/2017/01/15/shmoocon-2017-a-simple-tool-for-reverse-engineering-rf/.

       See also https://www.gnuradio.org/.

[25] Pasham, N., "RF Reverse Engineering has Become Trivial—Thanks to the 'Opensource SDR' Movement," *Medium*, Oct. 10, 2018, https://medium.com/@nihal.pasham/rf-reverse-engineering-has-become-trivial-thanks-to-the-opensource-sdr-movement-d1f9216f2f04.

       Julien, "Reverse Engineering a Wireless Home Alarm", PandwaRF, Aug. 15, 2019, https://pandwarf.com/news/reverse-engineering-a-wireless-home-alarm/.

[26] "Reverse Engineering Digital RF Signals the Easy Way with DSpectrum," Sep. 22, 2016, https://www.rtl-sdr.com/reverse-engineering-digital-rf-signals-the-easy-way-with-dspectrum/.

[27] U.S. Department of Health & Human Services, "Guidance Regarding Methods for De-identification of Protected Health Information in Accordance with the Health Insurance Portability and Accountability Act (HIPAA) Privacy Rule," Nov. 26, 2012, https://www.hhs.gov/hipaa/for-professionals/privacy/special-topics/de-identification/index.html#standard.

Glossary

The definitions provided below are generally harmonized with the NIST CSRC Glossary (https://csrc.nist.gov/glossary) and the U.S. CERT's NICCS Glossary (https://niccs.us-cert.gov/about-niccs/glossary).

AAMI   Association for the Advancement of Medical Instrumentation

Artifact   "A piece of evidence, such as text or a reference to a resource, that is submitted to support a response to a question [1]." Artifacts are generated for and maintained in the device's Design History File (DHF), and used to support such activities as regulatory submissions, postmarket surveillance, vulnerability management, and incident response. Artifacts may be physical or electronic in nature.

ASL   Approved supplier list

Attack surface   "The set of ways in which an adversary can [gain access] to a system and/or execute a threat [2]."

Attack vector   Also referred to as an attack path. "The steps that an adversary takes or may take to plan, prepare for, and execute an attack [2]."

Authentication   The process of "verifying the identity of a user, process, or device" [1].

Authorization   The process of "verifying that a requested action or service is approved for a specific entity [1]."

Availability   The ability to "provide timely, reliable access to information or a service [1]."

Business risk   Encompasses potential negative outcomes such as exposure of IP or trade secrets, loss of reputation or customer relationship and market share, legal repercussions, and financial losses. Excludes risks involving potential harm to device users. See also Risk.

CAPA   Corrective and preventive action

CERT   Computer Emergency Response Team, a division of the Carnegie Mellon Software Engineering Institute. CERT is registered trademark of the SEI. Other CERTs exist (in agreement or cooperation with CERT), (e.g., US-CERT is the national computer security incident response team (CSIRT) for the United States).

CERT/CC   CERT Coordination Center

CISA   Cybersecurity & Infrastructure Security Agency, a division of the United States Department of Homeland Security.

Compensating control   "A management, operational, and/or technical control (i.e., safeguard or countermeasure) employed by an organization in lieu of a recommended security control in the low, moderate, or high baselines that provides equivalent or comparable protection" [1]. Includes temporary security measures.

Confidentiality   The ability to "preserve authorized restrictions on information access and disclosure, including means for protecting personal privacy and proprietary information [1]."

COTS   Commercial off-the-shelf (software)

CSRC   Computer Security Resource Center, a division of NIST.

CVD   Cordinated vulnerability disclosure

Defense-in-depth   "The application of multiple countermeasures in a layered or stepwise manner to achieve security objectives. The methodology involves layering heterogeneous security technologies in the common attack vectors to ensure that attacks missed by one technology are caught by another [1]."

DFU   Device firmware update; see Update

DHS   The United States Department of Homeland Security

Disclosure, coordinated   Disclosure scenario where the Reporter informs the Coordinator, and either the Coordinator or the Reporter and Coordinator together informs the MDM, for example, in an ISAO.

Disclosure, full   Disclosure scenario where the Reporter publishes the analysis of the vulnerability as quickly as possible, making the analysis information accessible to everyone without delay or restriction.

Disclosure, non-   Disclosure scenario where the Reporter does not share the discovered cybersecurity information, except (A) only with the MDM under contract, bug bounty, or NDA; or (B) when shared with the intelligence community as a secret vulnerability for use in a future attack.

Disclosure, responsible Disclosure scenario where the Reporter informs the MDM of the vulnerability first, sharing the analysis with nobody besides the MDM until either: (A) the MDM releases remediation to address the vulnerability or initiates a voluntary recall (if warranted); or (B) an agreed-upon period of time has elapsed.

Efficacy The device's ability to provide clinically significant results when used as designed. See also 21 CFR Section 860.7(e)(1).

ENISA The European Union Agency for Cybersecurity (originally the European Network and Information Security Agency)

EOS End of support/end of service

EOL End of life

ETSI European Telecommunication Standards Institute

EU The European Union

Event An observable occurrence in an information system or network or attempt to "change the security state of the system (e.g., change access controls, change the security level of a user, change a user password). Also, any [occurrence] that attempts to violate the security policy of the system (e.g., too many logon attempts) [1]."

Exploit "A technique to breach the security of a network or information system in violation of security policy [2]."

Fallback mode A mode of operation with reduced functionality that maintains the essential performance of the medical device.

FDA The United States Food and Drug Administration

Fielded A fielded device or device in the field has been approved and issued for use by entities external to the MDM (e.g., for trials, demos, or active clinical or home health care use).

Forensics "The practice of gathering, retaining, and analyzing computer-related data for investigative purposes in a manner that maintains the integrity of the data [1]."

FOTA Firmware over the air.

Guidance Recommended approaches and methodologies for complying with preexisting law, often used in the context of providing additional interpretation and clarification of existing laws or regulations.

H-ISAC Health Information Sharing and Analysis Center

Hardening A process intended to eliminate a means of attack by reducing the attack surface and remediating vulnerabilities (e.g., by patching, removing or turning off unnecessary services, or closing unused ports).

HDO   Healthcare Delivery Organization

HIMSS   Health Information Management and Systems Society

HSCC   Health Sector Coordinating Council

HTA   Health Technology Alliance, a cooperative effort between HIMSS, AAMI, and ACCE

IDS   Intrusion detection system. Can be appended with H (Host IDS) or N (Network IDS).

IETF   Internet Engineering Task Force

IMDRF   International Medical Device Regulators Forum

Incident   "An occurrence that actually or potentially jeopardizes the confidentiality, integrity, or availability of a system or the information the system processes, stores, or transmits or that constitutes a violation or imminent threat of violation of security policies, security procedures, or acceptable use policies [1]."

Integrity   The ability to "guard against improper information modification or destruction, and includes ensuring information non-repudiation and authenticity [1]."

IPS   Intrusion prevention system. Can be appended with H (Host IPS) or N (Network IPS).

ISAO   Information Sharing & Analysis Organization

ISO   International Standards Organization

Legacy   Any device, component, or software that is not or cannot be brought up to current security standards or security baseline.

Lifecycle   The total activity timeline of a device, from initial concept exploration and component selection through end-of-service, decommissioning, or third-party resale acquisition. Encompasses multiple subcycles, including the supply chain lifecycle, development lifecycle, HDO lifecycle, and maintenance lifecycle.

Likelihood   "A weighted factor based on a subjective analysis of the probability that a given threat is capable of exploiting a given vulnerability [1]."

MDCG   Medical Device Coordination Group

MDIC   Medical Device Innovation Consortium

MDM   Medical device manufacturer

MDR   Medical Device Regulation

MDS²   Manufacturer Disclosure Statement for medical device security

MFA   Multifactor authentication; see Authentication

Mitigation Preventive measures intended to reduce the ease of exploitation and/or impact of a risk. Generally, the application of one or more measures to increase the difficulty of an unwanted occurrence and/or lessen its consequences. Implementing appropriate risk-reduction controls is based on risk management priorities and analysis of alternatives.

NIST National Institute of Standards and Technology, a division of the United States Department of Commerce.

NTIA The National Telecommunications and Information Administration, a division of the United States Department of Commerce.

OS Operating system

PHI Protected health information

PII Personally identifiable information

Postmarket The period of the device lifecycle that begins when the MDM places its device on the market and ends on the MDM's declared end-of-support date.

Postmarket surveillance Monitoring activities for possible threats, vulnerabilities, and incidents that potentially impact the device during its postmarket period.

Premarket The period of the device lifecycle that encompasses all activities up until the date the MDM has obtained market approval and places its device on the market.

Regulation/regulatory Referring to the legal context, including laws, policies, and overseeing bodies, that apply to a given medical device market.

Remediation "The act of mitigating a vulnerability or a threat [1]."

Reporter Anyone who discovers and attempts to disclose a potential threat or vulnerability.

Risk In cybersecurity, the potential for an unwanted or adverse outcome resulting from an incident, event, or occurrence, as determined by the level of difficulty for a particular threat to exploit a particular vulnerability and weighted using the exploit's associated consequences (i.e., severity).

Risk management "The process of identifying, analyzing, assessing, and communicating risk and accepting, avoiding, transferring or controlling it to an acceptable level considering associated costs and benefits of any actions taken. Includes (1) conducting a risk assessment, (2) implementing strategies to mitigate risks, (3) continuous monitoring of risk over time, and (4) documenting the overall risk management program [2]."

Root cause The vulnerability or vulnerabilities leveraged in a cybersecurity incident.

Safety   The determination that the probable health benefits from using the device outweigh any probable risks. See also 21 CFR Section 860.7(d)(1). Also defined as freedom from unacceptable risk; see also ISO/IEC 80001-1-1).

SAST   Static analysis security testing

SBOM   Software bill of materials

Security Systems Engineering   "Systems security engineering is a specialty engineering field strongly related to systems engineering. It applies scientific, engineering, and information assurance principles to deliver trustworthy systems that satisfy stakeholder requirements within their established risk tolerance" [1].

Severity   A vulnerability scoring metric intended to represent and rank the total potential impact of a given vulnerability being exploited by a threat.

SOUP   Software of unknown provenance, see TPSC.

Standard   "A published statement on a topic specifying the characteristics, usually measurable, that must be satisfied or achieved to comply with the standard [1]."

Supplier   "An organization or individual that enters into an agreement with the acquirer or integrator for the supply of a product or service. This includes all suppliers in the supply chain. Includes (1) developers or manufacturers of information systems, system components, or information system services, (2) vendors, and (3) product resellers" [1]. For purposes of this book, "agreement" as used in this definition includes general public licenses and open-source licenses as employed by TPSCs.

Testing, fuzz   A type of security testing in which "invalid data is input into the application via the environment, or input by one process into another process. Fuzz testing is implemented by tools called fuzzers, which are programs or scripts that submit some combination of inputs to the test target to reveal how it responds [1]."

Testing, penetration   A type of "security testing in which evaluators mimic real-world attacks in an attempt to identify ways to circumvent the security features of an application, system, or network. Penetration testing often involves issuing real attacks on real systems and data, using the same tools and techniques used by actual attackers. Most penetration tests involve looking for combinations of vulnerabilities on a single system or multiple systems that can be used to gain more access than could be achieved through a single vulnerability [1]."

Testing, Red Team   A Red Team is "a group of people authorized and organized to emulate a potential adversary's attack or exploitation capabilities against an enterprise's security posture. The Red Team's objective is to improve enterprise cybersecurity by demonstrating the impacts of successful attacks and by demonstrating what works for the defenders (i.e., the Blue Team) in an operational environment [1]."

Testing, static analysis A type of security testing that "detects software vulnerabilities by examining the app source code and binary and attempting to reason over all possible behaviors that might arise at runtime [1]."

Threat The potential for a threat-source (attacker or adversary) to exercise (accidentally trigger or intentionally exploit) a specific vulnerability.

Threat modeling "Threat modeling is intended to be a systematic and repeatable method of identifying cybersecurity threats that could exploit the weaknesses of a system [3]." See also Vulnerability assessment.

TPSC Third-party software component

V&V Verification and validation

Vulnerability "Weakness in a system, system security procedures, internal controls, or implementation that could be exercised by a threat" [1].

Vulnerability Assessment "Systematic examination of an information system or product to determine the adequacy of security measures, identify security deficiencies, provide data from which to predict the effectiveness of proposed security measures, and confirm the adequacy of such measures after implementation [1]."

Workaround See Compensating control.

References

[1] NIST CSRC Research Center Glossary, https://csrc.nist.gov/glossary.

[2] U.S. CERT NICCS Glossary, https://niccs.us-cert.gov/about-niccs/glossary.

[3] Leone, J., A. Wirth, V. Murthy, and T. Heyman, "Medical Device Threat Modeling," Med-Crypt, March 2020, p. 3.

About the Authors

Axel Wirth

Axel Wirth is the chief security strategist at MedCrypt. In this role he helps guide the company and its clients in critical security strategy decisions and assists in the adoption of leading security technology by the healthcare industry. As an advocate for compliance, privacy, and security—and ultimately patient safety—in healthcare, Wirth draws from over 30 years of international experience in the industry. He is an active participant in industry organizations, serves on boards and committees, and is a frequent speaker at conferences, forums, and webcasts on subjects such as healthcare cybersecurity and privacy, medical device security, regulatory compliance, and related healthcare-specific topics. In recognition of his accomplishments, he has been awarded the 2018 ACCE/HIMSS Excellence in Clinical Engineering & IT Synergies Award and the ACCE 2019 Clinical Engineering Advocacy Award, and has been recognized as a Fellow by the Association for the Advancement of Medical Instrumentation (AAMI) and Healthcare Information and Management Systems Society (HIMSS). His extensive background in the healthcare IT and medical device industries includes engineering leadership as well as strategic business development and marketing roles with Siemens Medical, Analogic, Mitra, Agfa Healthcare, and Symantec.

Christopher Gates

Christopher Gates is the principal systems security architect at Velentium. He has over 30 years of experience developing and securing medical devices, and works with numerous industry-leading device manufacturers and frequently collaborates with regulatory and standard bodies including the NTIA, MITRE, Bluetooth SIG, IEEE, the U.S. Department of Commerce, and the FDA to present, define, and codify tools, techniques, and processes that enable the creation of secure medical devices. Everywhere he goes, Gates promotes the secure development lifecycle described in this book, believing that this approach ultimately eases the burden on developers and ensures high-quality products that can work as intended to save and improve lives without putting users at risk. This motivation became tangible to Gates when his first grandchild was born prematurely, before his lungs could fully close. When Gates visited him in the NICU, he noticed that a handful of the machines allowing his grandson to breathe had been designed by Gates himself.

Jason Smith

Jason Smith is a marketing strategist and technical writer at Velentium, where he supports engineering teams and clients developing quality documents, user manuals, DHF and DMR templates and content, and submissions artifacts. Smith has been working as a professional writer in various fields and capacities for over 20 years, and was asked to apply his skill set to medical devices beginning in 2016. *Medical Device Cybersecurity for Engineers and Manufacturers* is the 8th published book he has helped author.

Contributors

The authors would like to gratefully acknowledge the contributions of medical device cybersecurity leaders from both the public and private sectors who lent their expertise in key subjects to this book:

Alan Friedman, for the SBOM material in Chapter 8. Dr. Friedman is Director of Cybersecurity Initiatives at National Telecommunications and Information Administration (NTIA) in the U.S. Department of Commerce. He coordinates NTIA's multistakeholder processes on information security, focusing on addressing vulnerabilities across the software world. Prior to joining government service, Dr. Friedman was a noted cybersecurity and technology policy researcher at Harvard's Computer Science Department, the Brookings Institution, and George Washington University's Engineering School. He is coauthor of the popular text *Cybersecurity and Cyberwar: What Everyone Needs to Know*.

Greg Garcia, for the Foreword. Garcia is Executive Director for the Health Sector Coordinating Council's Cybersecurity Working Group and the nation's first Assistant Secretary of Cybersecurity and Communications for the U.S. Department of Homeland Security from 2006–2009.

Michelle Jump, for the regulatory and standards material in Chapter 2. Jump is Global Regulatory Advisor for Medical Device Cybersecurity at MedSec, where she is responsible for providing strategic leadership, training, and advisory services to the medical device industry in the area of cybersecurity compliance, global regulations, standards, product security program development, and security risk management. She actively participates in a variety of domestic and international standards committees and working groups, as well as relevant industry and governmental initiatives to support security within the health care industry.

Satyajit "Sat" Ketkar, for the SAST and unit testing material in Chapter 8. Ketkar is the principal systems engineer and systems architect at Velentium. His 20-year career includes electrical, firmware, software, and systems engineering experience, including seven in medical device design and two with a European Union notified body conducting medical device reviews and audits for safety, quality, performance, and security.

Michael McNeil, for the security governance and building a security-capable organization material in Chapter 2. McNeil is the SVP and Global CISO at McKesson. In his previous role as global product security & services officer at Philips, he developed the company's product security program, ensuring consistent processes leading to an improved device security posture. Philips' Security Center of Excellence was awarded the first-ever UL IEC 62304 certification from Underwriters Laboratories in March 2020. McNeil has over 20 years of security and privacy expertise.

Eric Pancoast, for the cryptography and high resource bandwidth material in Chapter 10. Pancoast currently serves as CTO at MedCrypt, a company focused on building proactive security solutions for medical devices. Pancoast was cofounder of Gamma Basics, a radiation oncology focused software startup that was acquired by Varian Medical Systems in 2013. In an earlier role at Mimeya Technology, Pancoast served as lead architect and software engineer.

Index

For further information on these and other Artech House titles, including previously considered out-of-print books now available through our In-Print-Forever® (IPF®) program, contact:

Artech House
685 Canton Street
Norwood, MA 02062
Phone: 781-769-9750
Fax: 781-769-6334
e-mail: artech@artechhouse.com

Artech House
16 Sussex Street
London SW1V 4RW UK
Phone: +44 (0)20 7596-8750
Fax: +44 (0)20 7630-0166
e-mail: artech-uk@artechhouse.com

Find us on the World Wide Web at: www.artechhouse.com